Jean Bosco IGIRUKWISHAKA
Frédéric BANGIRINAMA

Characterization of invasive marsh plants

Jean Bosco IGIRUKWISHAKA
Frédéric BANGIRINAMA

Characterization of invasive marsh plants

Plants in the marshes of the Ruvubu River, one of Burundi's four Ramsar sites

ScienciaScripts

Imprint

Cover image: www.ingimage.com

This book is a translation from the original published under ISBN 978-620-6-70222-1.

Publisher:
Sciencia Scripts
is a trademark of
Dodo Books Indian Ocean Ltd. and OmniScriptum S.R.L publishing group

120 High Road, East Finchley, London, N2 9ED, United Kingdom
Str. Armeneasca 28/1, office 1, Chisinau MD-2012, Republic of Moldova, Europe
Printed at: see last page
ISBN: 978-620-7-61611-4

JURY COMPOSITION

Chairman : Prof. NDAYISHIMIYE Joël

Secretary: Dr. BARARUNYERETSE Prudence

Director: Prof. BANGIRINAMA Frédéric

DEDICACES

To my late father BAMPUHIYE Fabien

To my mother BANGIRINAMA Rosette

To my wife NIYUNGEKO Marie Jeanine

To our children IGIRUKWISHAKA Shana Lenaëlle and IGIRUKWISHAKA Chély Asaël

ACKNOWLEDGEMENTS

At the end of this work, I would like to express my deep gratitude to all those who, in one way or another, have contributed to its realization.

I would particularly like to thank Professor BANGIRINAMA Frédéric, the supervisor of this thesis, who, despite his many responsibilities, was kind enough to supervise it. His scientific vision, skills, unrivalled dedication, advice and pertinent comments were of great use to me. May he be assured of my deepest gratitude.

I would also like to thank the chairman of the jury, Professor Joël NDAYISHIMIYE, and Dr. Prudence BARARUNYERETSE, secretary to the jury, for agreeing to devote part of their precious time to reading, correcting and improving this work. Many thanks for your constructive and judicious comments.

I would also like to express my gratitude to all the teachers in the Master's program in integrated environmental science and management at the University of Burundi. Their scientific and human training enabled me to bring this work to fruition.

My gratitude also goes to my dear parents, who not only showed me the way to school, but also supported my education from elementary school to university. I would like to express my deepest gratitude to them.

Allow me to make a very warm special mention of my dear wife NIYUNGEKO Marie Jeanine: sincere gratitude for your patience and unfailing support. You single-handedly supported Shana Lenaëlle and Chély Asaël during these two years, without counting on my responsibility. This memoir is yours.

Last but not least, I'd like to express my gratitude to everyone who has contributed in any way to the realization of this work, and to all my friends who have given me both moral and material support.

SUMMARY

This work involved studying the flora of the marshes of the Ruvubu, one of Burundi's most important rivers. Its overall aim was to contribute to the knowledge of invasive plants present in the marshes of the Ruvubu river, with a view to providing useful knowledge for environmental managers. Composition and diversity were analyzed by establishing species richness, calculating diversity indices and studying species traits. Structure was analyzed by identifying plant groupings. A systematic approach based on phytosociological surveys was used to establish the floristic richness of the river marshes. The identification of invasive plants was based on a combination of the average cover rate and the biological characteristics of the invasive specific plasticity of the dominant species. The results show that the flora is diverse, with 161 species in 54 families and 125 genera. The families with the highest number of species are Poaceae (17.39%), Asteraceae (14.91%), Fabaceae (10.56%), Cyperaceae (4.97%), Malvaceae (3.73%) and Euphorbiaceae (3.11%). The weighted spectrum revealed the dominance of champhytes (33.73%), anemochores (23.49%) and widespread species (62.04%). Four plant groupings were identified according to the Bray-Curtis species dissimilarity index: grouping G1 with 13 records (83 species), grouping G2 with 9 records (67 species), grouping G3 with 6 records (77 species) and grouping G4 with 10 records (92 species). The combination of average species cover and invasive plasticity characteristics resulted in the selection of 13 proven invasive species and 7 other potential invasive plant species. The invasive plants selected fall into six botanical families, two of which (Poaceae and Asteraceae) are the most represented, accounting for 69.23% of invasive species. The floristic gradient of selected invasive plants increases from upstream to downstream.

Keywords: Marais de la rivière Ruvubu, Diversity, invasive plants, Burundi

ABSTRACT

This research consisted in studying the flora of the Ruvubu marshes, one of the most important rivers in Burundi. The overall objective of this study was to contribute to the knowledge of invasive plants present in the marshes of the Ruvubu River in order to provide useful knowledge to environmental managers. Composition and diversity were analyzed by establishing species richness, calculating diversity indices and studying the species' life traits. The structure was analyzed through the identification of plant groups. The method used is the systematic approach based on phytosociological surveys to establish the floristic richness of the river marshes. The determination of invasive plants required a combination of the average cover rate and the biological characteristics of the invasive specificity of the dominant species. The results show that the flora is diverse with 161 species distributed in 54 families and 125 genera. The most represented families in terms of species are Poaceae (17.39%), Asteraceae (14.91%), Fabaceae (10.56%), Cyperaceae (4.97%), Malvaceae (3.73%) and Euphorbiaceae with 3.11% of inventoried species. The weighted spectrum showed the dominance of chamephytes (33.73%), anemochores (23.49%) and widespread species (62.04%). Four plant groups were identified based on species frequencies: group G1 with 13 records (83 species), group G2 with 9 records (67 species), group G3 with 6 records (77 species) and group G4 with 10 records (92 species). The combination of the average species cover rate with the characteristics of invasive plasticity retained 13 proven invasive species and 7 other potential invasive plant species. The invasive plants selected are distributed in six botanical families, two of which (Poaceae and Asteraceae) are the most represented, accounting for 69.23% of the invasive species. The floristic gradient of the selected invasive plants increases from upstream to downstream of the river.

Keywords: Ruvubu river marshes, Diversity, invasive plants, **Burundi**

TABLE OF CONTENTS

LIST OF ACRONYMS AND ABBREVIATIONS

APRN/BEPB	Association Protection des Ressources Naturelles pour le Bien-Etre de la Population au Burundi
GEE	Invasive Species Group
MEEATU	Ministry of Water, the Environment, Spatial Planning and Urban Development
SNAPB-2013-2020	National Biodiversity Strategy and Action Plan 2013-2020
IUCN/PACO	International Union for Conservation of Nature/Central and West Africa Program

FOREWORD

For mankind, the environment is a potential asset, a partner that needs to be better preserved and managed. To achieve this, we need a better understanding of its origins and mechanisms, its ecological role and its economic interest for mankind. Protecting the environment requires the effective participation of every institution. And for every citizen to participate effectively, basic knowledge must be acquired.

With this in mind, the University of Burundi has set itself the general objective of contributing to the rational management and protection of the environment at national, regional and global level.

To this end, it has opened a Master's program in Science and Integrated Environmental Management, which trains graduates in the management of space and natural ecosystems for optimal, sustainable use, as well as in the prevention and/or mitigation of the effects of various nuisances and pollution in both urban and rural environments.

This thesis entitled "**Characterization of invasive plants in the marshes of the Ruvubu River**" is part of the Master's degree mentioned above. The aim of this work is to gain knowledge of the invasive plants present in the marshes of the Ruvubu river. The aim is to contribute to the sustainable management of the Ruvubu National Park, which is so important at national, regional and international level. The idea for this research project arose from the fact that knowledge of invasive taxa, their level of proliferation and the areas affected is still lacking in Burundi.

This study is intended to contribute to knowledge of the biodiversity of the Ruvubu river marshes, and particularly their degree of invasion by exotic plants. This will guide the adoption of appropriate measures for their sustainable management.

INTRODUCTION

1. Issues

We're living in a time when biodiversity is disappearing at an unprecedented rate for a variety of reasons (Teyssèdre & Couvet, 2010); Primack et al., 2012; Mendoza, 2016; Ben-ghabrit et al., 2018. Ranking the causes of this degradation shows that, after habitat destruction, biological invasions are the second biggest source of destruction to terrestrial and aquatic ecosystems (Ben-ghabrit et al., 2018; IUCN/PACO, 2013)..

Patrick (2017) defines biological invasion as "a *process of geographic range extension of a taxon moving from one region to another, capable of reproducing in the new region and developing perennial populations there*".

Invasive plants cause damage or have the potential to destroy the environment, health or agricultural production. The many impacts caused by invasive plants are mainly related to their capacities for adaptation, reproduction and dispersal, which give them an advantage over native species (Abram & Moffat, 2018; Akodéwou et al., 2019). They can modify ecosystem functioning and the characteristics of the abiotic environment (Osawa et al., 2013).

Invasive plants are causing a new, more homogeneous distribution of species. This new distribution is so important that some scientists speak of a new evolutionary era called the "Homeogeocene" (Jacques et al., 2006; Jean-François et al., 2012).. However, biotic homogenization has a negative impact on the ability of communities to provide multiple ecosystem services. This is because not all species provide the same services with the same intensity (Van Der Plas et al., 2016).

Invasive plants are currently the focus of concern due to the threat they pose to biodiversity and ecosystem integrity (Primack et al., 2012; Nzigidahera & Habonimana, 2015; OSS, 2020). Their alarming proliferation has raised the awareness of a number of environmental protection stakeholders, with the result that their prevention and management has become one of the 20 Aïchi targets to be achieved by 2020, adopted by the States Parties to the Convention on Biological Diversity, including Burundi.

With a view to implementing this Aïchi objective, Burundi has set itself, through its National Strategy and Action Plan on Biodiversity 2013-2020, Objective 10: "*By 2015, the extents of invasive alien species and their pathways of introduction are identified, practical measures and appropriate legislation are put in place to control and eradicate the most dangerous species."* (MEEATU, 2013). Nevertheless, knowledge of invasive taxa, their level of proliferation and the areas affected remains lacking in Burundi. This is due to the fact that the studies on invasive plants and the map of their distribution targeted as indicators in objective 10 of SNAPB-2013-2020 have not been fully produced. This illustrates that Burundi's biodiversity manager still lacks reference tools for preventive and curative control of invasive plants.

A few studies on invasive plants have already been carried out in Burundi (Habonayo et al., 2019; Nzigidahera, 2017).. However, it has been noted that protected areas, especially the Rusizi and Kibira national parks, the Bururi reserve and the Bugesera protected aquatic landscape, have been targeted more than the rest of the country. The Imbo plain, particularly along the coast and the tributaries of Lake Tanganyika, has been the focus of a number of briefs focusing on this issue.

Wetlands appear to be particularly vulnerable to biological invasions, especially plant invasions (Joy & Kercher, 2004). Yet these areas, with their relative biodiversities, are particularly interesting insofar as they present particular ecological functions and socio-economic functions (MEEATU, 2014). The marshes that make up Burundi's wetlands occupy 117,993 hectares, or around 4.24% of the national territory, and are currently much disturbed by anthropogenic activities, especially agriculture (Bizuru, 2005). In addition, these disturbed areas play an important role in the invasion of exotic plants (GEE, 2011). They play a key role in the flow of invasive species between different landscape compartments. Depending on the practices of their managers, these spaces are either potential sources of propagation or, on the contrary, structuring elements limiting the expansion of certain invasive species (Gérard et al., 1998; Tassin, 2002). It is therefore necessary, as part of a systemic approach, to integrate them into the analysis of plant invasions in natural environments.

2. Subject interest

Burundi's marshland ecosystems remain poorly understood as a whole (Bizuru, 2005) and the marshes of the Ruvubu River are no exception. This work is a contribution to the knowledge of invasive species in Burundi. It is a study carried out in the least-surveyed or unsurveyed areas of Burundi (Masabo & Nindorera, 2019) including the marshes of the Ruvubu River. The latter drains more than a quarter of Burundi's water and is the southernmost tributary of the

River Nile. It rises in the mountains of the Congo-Nile ridge and forms part of the Upper Akagera basin. The Ruvubu River flows through a national park bearing its name, and is listed as one of Burundi's four Ramsar sites. This meandering river has a low flow between the entrance and exit of the park. The Ruvubu marshes are largely flooded and occupied by permanent swamps (MEEATU, 2014). Contributing to knowledge of its biodiversity and particularly its degree of invasiveness by exotic plants would guide the taking of good measures for its sustainable management.

3. Work objectives

The overall aim of the present research is to contribute to the knowledge of invasive plants present in the marshes of the Ruvubu river. The aim is to contribute to the sustainable management of the Ruvubu National Park, which is so important at national, regional and international level.

Specifically, the work consists of: (i) carry out a floristic inventory, i.e. the scientific identification and botanical description of the plants inventoried; (ii) highlight the diversity of invasive plants at the surveyed site and, finally, (iii) determine a floristic gradient of invasive plants from upstream to downstream.

4. Assumptions

The aim of this study is to test the hypothesis that the marshes of the Ruvubu River, being disturbed by agriculture, are home to a high level of plant diversity. Also, as the flow of the Ruvubu River is low (MEEATU, 2014) and the Ruvubu being crossed by some twenty bridges, the result would be a high spread and floristic abundance of invasive plants in its marshes. Invasive species in the marshes of this river would have a strong potential to spread to Ruvubu National Park from upstream to downstream.

5. Work delimitation

Although biological invasions concern all groups of organisms, from unicellular to multicellular, it is not possible to study all these groups at the same time. The present study focuses on the plants of the Ruvubu river marshes. This study is intended to be twofold: firstly, we have carried out a floristic analysis of the marshes of the Ruvubu River; secondly, our study addresses the identification of invasive plant species in the marshes of the Ruvubu River.

Alongside a general introduction, conclusions and suggestions, the work is divided into four chapters. The first chapter, "Generalities", deals with invasive plants and their harmful effects on ecosystems and their biodiversity. The second chapter, entitled "Materials and methods", describes the study site, the materials used and the methodological approach adopted to arrive at the results presented in the third. The latter describes the floristic composition of the environment surveyed and the characterization of the invasive species selected. The final chapter is devoted to a discussion of the results.

CHAPTER I: GENERAL INFORMATION ON BIOLOGICAL INVASIONS

The development of human communities has contributed to the breaking down of natural barriers and the expansion of species outside their original areas of distribution. Globalization, increased international trade and the resulting disruption of ecosystems have facilitated the introduction and dispersal of numerous animal and plant species. Some of these species have become harmful to the environment (Allen et al., 2017)and the phenomenon of biological invasion is on the increase.

Biological invasions have long been a well-known phenomenon (Richardson & Pyšek, 2007).. They are currently one of the major challenges facing natural and semi-natural ecosystems worldwide in the 21st^{ème} century. Losses and damage caused by invasive species have been steadily increasing over the years (Ben-ghabrit et al., 2018). Key food plants, trees and shrubs suitable for nesting and refuge for wild animals, plants that purify water and serve as symbionts for others, those that provide support for climbers and shelter delicate vegetation may also be compromised or even set to disappear due to invasive species (IUCN-PAPACO, 2013). Knowledge of these species is essential for their proper management, in order to achieve better conservation of biological diversity.

I.1 Definitions

I.1.1. Native and exotic species

I.1.1.1. Native species

Species are distributed according to their natural ranges. Ecologically speaking, species that occur naturally in a region without human intervention are said to be indigenous or autochthonous to that region. And each species is therefore qualified as indigenous (autochthonous) to a specific region. According to Toussaint et al. (2007)for a given biotope, plants are considered indigenous when they were widespread in a region before the year 1500. These plants constitute a species or population native to a given area, as opposed to introduced species known as allochtones.

I.1.1.2. Exotic (allochthonous) species

For many years now, humans have been traveling from one region to another, and even from one continent to another. These journeys are often accompanied by the voluntary or involuntary introduction of foreign species into a given biogeographical region. These newly-introduced species are known as exotic or allochthonous species.

According to Lisan (2014)exotic species are likely to be introduced and spread, voluntarily or accidentally, by various routes, including :

- horticulture and gardening;
- seed mixtures ;
- increased national, regional and international trade and travel;
- artificial canals ;
- pleasure and commercial boating ;
- containers and vessels in the maritime fleet;
- unauthorized introductions; etc.

I.1.2 Biological invasions

The definitions of biological invasion are numerous and vary according to the authors, who may or may not include notions of the ecological and/or economic impacts of these organisms (Julie et al., 2007; Soubeyran, 2010; Van Kleunen et al., 2010; Rejmánek et al., 2013).

Moreover, there is still a lack of consensus as to the origin (native or exotic) of an invasive taxon. Some studies speak of biological invasion when a non-native taxon is introduced into a new environment (ecosystem or habitat) and spreads, causing damage to native biodiversity under conservation (Jean-François et al., 2012; IUCN-PAPACO, 2013; Ben-ghabrit et al., 2018). For this to happen, a taxon that is not represented in the vegetation of an area must enter from "outside", survive and reproduce, spread from its point of introduction, naturalize and spread further - ultimately causing damage. Yet other recent studies show the invasion of natural ecosystems by known indigenous species (Habonayo et al., 2019; Lisan, 2014; Nzigidahera, 2017; Zihalirwa et al., 2020).

In this work, the definition that reconciles the two considerations has been adopted, as proposed by Valéry et al (2008). These authors define an invasive taxon as an exotic or native taxon that has a competitive advantage enabling it, following the disappearance of natural barriers to its proliferation, to spread rapidly and dominate new areas in recipient ecosystems, within which it becomes a dominant population.

In this case, we consider an invasive plant to be any native or non-native plant with the ability to colonize an area rapidly, spread far from parent plants and cause or be likely to cause economic or environmental damage, or harm to human health.

I.1.3. Invasive species

The term "invasive" is used to characterize a taxon with a high capacity for proliferation, whether exotic or indigenous to a given territory. Invasive species are opportunistic, with a high capacity for development and competition. As a result, they pose a threat to a biotope and its biocenosis, which can disappear in extreme cases.

I.2 Types of invasion

I.2.1. Native species and invasiveness

It is rare for a native species to become invasive in its natural environment. However, with the repetitive disturbances that an ecosystem may undergo, a species or group of species may find conditions favorable to its flourishing and then show a dynamic of rapid extension in its native range to the detriment of others (Canton du Valais, 2017; Zihalirwa et al., 2020). These species can remain in a non-invasive state for decades (or even centuries) and then start to spread and cause damage inside and outside the biotope under consideration. This delay is sometimes referred to as the "latency phase" of invasion, and may be due to a number of factors.

The invasive nature of a species may be latent due to the existence of a biological regulator that uses the organs or individuals of this species to ensure its vital functions, such as a predator. (Masumbuko, 2011). This latent phase of invasion may also be due to the slow adaptation of a (potentially invasive) species to its environment before a viable and dispersible seed is produced in sufficient quantity to begin the propagation and subsequent stages that will cause damage to biodiversity (UICN/PACO, 2013). A species that is not normally invasive can become so if a factor is present that favors its competition to the detriment of others (Dukes & Mooney, 1999).

I.2.2. Exotic species and invasiveness

The invasion process, and its stages from introduction (for exotic taxa) to invasion, can last for years or even decades or centuries (UICN/PACO, 2013). A plant may not be invasive for a certain period of time, and then at some point fall into the invasive alien category; the reverse is also possible (Agboola & Joseph, 2014). That's why it's ideal to notice new species that arrive (exotic species) and mix with the flora of an ecosystem, and to check whether elsewhere they have a reputation as invasive species.

Among exotic species, invasive plants are those with the greatest impact on biodiversity. In fact, not all non-native plants become invasive. In fact, they represent only a very small proportion of introduced species. According to the "rule of threes" evoked by Fourdrigniez & Meyer (2008)it is estimated that out of 1000 introduced species, 100 manage to acclimatize, 10 naturalize (i.e. reproduce and disperse over long distances without human intervention) and only one becomes invasive, i.e. with a significant impact on biodiversity.

Quéré et al . (2011) distinguish invasive alien species into three groups: proven invaders, potential invaders and invaders to watch out for.

I.2.2.1. Proven invasive species (IA)

These are non-native plants which, in their territories of introduction, have a proven invasive character and have a negative impact on biodiversity and/or human health and/or economic activities. In this group, Bousquet et al. (2016) recognize three categories:

- **category IA1**: naturalized plants or plants in the process of naturalization which are currently invasive within natural or semi-natural plant communities in the area under consideration, and which compete with native species or produce significant changes in the composition, structure and/or functioning of ecosystems (these are known as transforming species);
- **category IA2**: naturalized plants, or those in the process of becoming naturalized, that are currently known to be invasive in the area under consideration, in natural or semi-natural environments, or in highly anthropized environments (farms, roadsides, etc.), and are causing serious problems for human health.
- **category IA3**: naturalized plants or plants in the process of becoming naturalized which are currently invasive within natural or semi-natural plant communities in the area under consideration, and which cause damage to certain economic activities.

I.2.2.2. Potential invasive plants (IP)

These are non-native plants currently showing a tendency to develop an invasive character within natural or semi-natural communities, and whose dynamics within the territory under consideration and/or in neighbouring or climatically close regions, are such that there is a risk of it becoming an established species in the more or less long term. As such, the presence of potential invasive species in the territory under consideration warrants a high level of vigilance, and may require the rapid implementation of preventive or curative actions. Potential invasive species include the following categories according to Quéré et al (2011) :

- **IP1 category**: plants absent from the territory under consideration, but determined to be invasive in a directly adjacent territory and which present a risk of appearing in the near future due to their extension dynamics;
- **IP2 category**: naturalized plants or plants in the process of naturalization currently showing, in the territory under consideration, a proven invasive character only within highly anthropized plant communities (farms, roadsides, etc.), and which show an invasive character (with impact on local biodiversity) within natural or semi-natural plant communities elsewhere in the world in a close climatic range;
- **IP3 category**: accidental plants, naturalized or in the process of naturalization, which currently show a tendency to develop an invasive character in natural or semi-natural environments, or in highly anthropized environments, and which cause serious problems for human health;
- **IP4 category**: accidental plants showing a tendency to develop an invasive character within natural or semi-natural plant communities in the territory under consideration, and which show an invasive character (with impact on local biodiversity) within natural or semi-natural plant communities elsewhere in the world in a close climatic range.
- **IP5 category**: naturalized plants or plants in the process of becoming naturalized which, in the area under consideration, show a tendency to develop an invasive character within natural or semi-natural plant communities and seem likely to harm local biodiversity.

I.2.2.3. Plants to watch (AS)

In natural or semi-natural environments, Quéré et al. (2011) define a "plant to watch" as any non-native plant that does not currently (or no longer) have a proven invasive character or negative impact on biodiversity in the area under consideration, but for which the possibility of developing these characteristics (through sexual reproduction or vegetative propagation) has not been completely ruled out. They propose to take into account the invasive nature of this plant and its impact on biodiversity in other regions. The presence of such plants on the territory under consideration, in natural or anthropized environments, requires special monitoring, and may justify rapid intervention measures.

According to Bousquet et al (2016)the following exogenous plants are included among the plants to watch out for:

- **AS category**: accidental, naturalized or naturalizing plants that do not currently show a tendency to develop an invasive character in the territory under consideration (no dense population development in at least one site, nor rapid extension dynamics) in natural or semi-natural environments, or in highly anthropized environments (farms, roadsides, etc.), but which are known to cause serious problems to human health.
- **AS2 category**: naturalized plants or plants in the process of naturalization that are currently invasive only within highly anthropized plant communities in the area under consideration, but are not considered invasive within natural or semi-natural plant communities elsewhere in the world in a similar climatic zone.
- **AS3 category**: accidental plants showing a tendency to develop an invasive character within natural or semi-natural plant communities in the territory under consideration, and not being considered invasive within natural or semi-natural plant communities elsewhere in the world in a close climatic range.
- AS4 category: accidental plants, naturalized or in the process of naturalization in natural or semi-natural environments, or in highly anthropized environments, which do not currently show a tendency to develop an invasive character (no development in dense population in at least one site, nor rapid extension dynamics) in the territory under consideration, but which have shown in the past an invasive character (with impact on biodiversity) in the territory under consideration, and are now integrated without dysfunction into native communities.
- **AS5 category**: accidental, naturalized or naturalizing plants that do not (or no longer) currently show a tendency to develop an invasive character in the area under consideration (no dense population development in at least one site, nor rapid extension dynamics), but are considered to be proven invaders (invasives with an impact on

biodiversity) elsewhere in the world in a nearby climatic area, within natural or semi-natural plant communities.

- **category AS6**: accidental plants, naturalized or in the process of naturalization, presenting in the territory under consideration a tendency to develop an invasive character within plant communities strongly influenced by man, and being considered invasive (and damaging to local biodiversity) elsewhere in the world in a nearby climatic area, within natural or semi-natural plant communities.

These three groups of invasive alien species (proven invasive species, potential invasive species and invasive species to watch out for) are summarized in a table by Bousquet et al, (2013).

Table I.1. Summary: classification of invasive plants as "proven invasives", "potential invasives" and "plants to watch out for".

Situation of the plant in the area under consideration	**Plant category**	
Exogenous plant not reported on the territory but - considered invasive in a neighbouring department - not considered invasive in an adjacent territory	→Potential invasiveness →Non-invasive	IP1 -
Native plant (even if locally proliferating)	→Non-invasive	-
Exogenous plant causing serious problems for human health - of proven invasiveness - with a tendency to be intrusive - no tendency to develop an invasive character	→Proven invasive →Potential invasiveness →Watch out	IA2 IP3 AS1
Exogenous plant with a proven invasive character in a natural or semi-natural environment and - harming biodiversity or - causing problems for economic activities	→Proven invasive →Proven invasive	IA1 IA3
Exogenous plant invasive only in environments strongly influenced by man (embankments, rubble, etc.): - if an impact on biodiversity is known in natural environments in other regions of the world (with a similar climate) - if there is no known impact on biodiversity in natural environments in other regions of the world (with a similar climate)	→Potential invasiveness →Watch out	IP2 AS2
Exogenous plant with a tendency to become invasive only in environments strongly influenced by man (embankments, rubble, etc.):	→Watch out	AS6

- if an impact on biodiversity is known in natural environments in other regions of the world (with a similar climate) - if there is no known impact on biodiversity in natural environments in other regions of the world (with a similar climate)	→Non-invasive (*no a priori risk for natural environments*)	-
An exogenous plant with a tendency to become invasive in a natural or semi-natural environment: - Naturalized or naturalizing plant - Accidental plant (recent establishment, not stabilized) : - if an impact on biodiversity is known in natural environments in other regions of the world (with a similar climate) - if there is no impact on biodiversity	→Potential invasiveness →Potential invasiveness →Watch out	IP5 IP4 AS3
Plant that is not (or is no longer) invasive: - whether the plant has been classified in the past as invasive in the natural environment - if the plant has not previously been classified as invasive and : - if an impact on biodiversity is known in natural environments in other regions of the world (with a similar climate) - if there is no known impact on biodiversity in natural environments in other regions of the world (with a similar climate)	→Watch out →Watch out →Non-invasive	AS4 AS5 -

Source: Table found in "Liste des plantes vasculaires invasives de Basse-Normandie" by (Bousquet et al., 2013)

I.3 Ecological characteristics of invasive species

Many authors have attempted to define the typical profile of invasive species (Fumanal, 2007). Invasive species succeed in colonizing new ecosystems thanks to a number of characteristics that make them more competitive than local species. The same characteristics make them more difficult to control. Their greater ecological amplitude also favors their development in highly diversified biotopes (Fumanal, 2007; Fourdrigniez & Meyer, 2008). Thus, UICN/PACO (2013) notes that species that become invasive converge on the following points :

- a rapid growth rate that exceeds that of native plants,
- remarkable expansion characteristics, enabling propagules to spread rapidly and widely,
- high reproductive capacity, often producing large quantities of seeds or other propagules,
- high environmental tolerance, whereas native species often exist within narrow limits of temperature, rainfall, soil type, etc,

- mechanisms that make them effective competitors with local species for water, nutrients, light and space in which to develop,
- production of allelopathic substances (by leaves, stems or roots) that prevent other species from germinating, growing or reproducing fully,
- absence or limited number of predators or other natural enemies.

Although the above properties help define invasive organisms, there are always exceptions. What's more, no species possesses all the typical characteristics of a good invader, and not all of these are essential for a species to become invasive (Vanderhoeven et al., 2006; Soubeyran, 2008).

I.4. Pathways for the introduction of invasive species

For centuries, man has intervened in nature, transporting thousands of species far from their areas of origin. But since the mid-20ème century, with the globalization of the economy and the accompanying development of transport, trade and tourism, species movements and biological invasions have accelerated considerably. (Vanderhoeven et al., 2006; Van Kleunen et al., 2010).. Introduction routes fall into two categories (Fumanal, 2007; Soubeyran, 2008; IUCN/PACO, 2013).

The first is accidental introduction into degraded or unoccupied areas where they can easily establish and spread once a pioneer population has become established. Such areas as roads, roadsides, railroads, airstrips, quarries, construction sites, drains, streams and even formal park entrances and parking areas can all bring propagules to sites where they can begin to establish plant populations in the absence of any competition.

The second common pathway for the introduction of invasive plants is the intentional planting of exotic species for production forestry, boundary marking, shade, beautification and even food production. These can be grasses, shrubs, garden plants or trees which, after a period of time, acclimatize and then become capable of spreading particularly if they have (or regain through gradual adaptation to their new habitat) one or more of the invasive characteristics. These may be species that are benign (and non-invasive) in other situations where they have natural enemies, but in a new locality are able to express their invasive tendencies. Or, for some flowering plants, it may take decades before a pollinator begins to visit the flowers and fertile seeds are produced.

Other authors such as Joy B. & Kercher (2004); Muoghalu & Chuba (2005) Blanfort et al. (2009) and Fleriag (2009) refer to several other pathways and vectors for the entry of exotic plants into a territory, such as people and their clothing, luggage, trade goods, deliveries, containers, construction materials, garbage and garden waste disposal, livestock movements, wildlife migrations, rivers and natural events such as storms and floods.

I.5. Success factors influencing biological invasion

There is no consensus on the factors behind the success of invasive species, nor on the possibility of predicting these invasions (Heger & Trepl, 2003; Guo, 2006; Williamson, 2006). The identification of characteristics that can predict which species are likely to become invasive, or which communities are likely to be invaded, has been the subject of much research and is still widely debated (Bifolchi, 2007). The probability of a species becoming invasive seems to result from the combination of multiple ecological, demographic and genetic factors. These are grouped into biotic and abiotic factors by Guo (2006) who stressed their importance in limiting the populations of biological invaders.

I.5.1. Biotic factors

The ability of a species to invade an ecosystem depends on its degree of adaptation to that environment, or its adaptive potential (Lee, 2002). The ecological plasticity of certain species (size, diet, habitat), by favoring the survival and reproduction of introduced individuals, is closely correlated with the success of their invasion (Cassey, 2002; Cassey et al., 2004; Fournier, 2018; Zihalirwa et al., 2020). The high adaptability to an environment and the ecological plasticity of invasive species constitute their aptitude for invasion and their first weapon for establishment in a new environment.

Demographic factors intrinsic to the species (growth rate, spread) and factors, often unique, associated with each introduction event, such as the number of individuals introduced and the frequency of introduction, can influence the behavior of a potentially invasive species. Numerous studies underline the pre-eminent role of introduction pressure, i.e. the number of individuals introduced as well as the frequency of introductions, in the success of biological invasions (Woods, 1997; Lefeuvre, 2016).

The characteristics of the host community also influence the success of an introduced species' invasion. Interactions between introduced and native species can induce evolutionary changes through competitive exclusion, niche displacement, hybridization, predation and even extinction (Woods, 1997; Guo, 2006; Mooney & Cleland, 2001; Williamson, 2006). However, the evolutionary aspects of invasions are little studied (Lee, 2002) so the fact that invasions involve the introduction of new genotypes and alleles is downplayed (Piry et al., 2004).

In addition, human activities are very likely to generate settlement opportunities by creating new resources, reducing the presence of natural enemies (Zihalirwa et al., 2020)etc. According to Vitousek et al. (1997) and Rejmánek et al. (2013)invasive events can be favored in disturbed environments and degraded habitats. Disturbances make environments susceptible to invasion by creating openings in which the likelihood of a new species becoming established is generally higher. Conversely, natural environments that are little disturbed and well structured are likely to be highly resistant to invasion.

I.5.2. Abiotic factors

Invasive species evolve in response to their environment. The similarity and proximity of areas of origin and introduction play a major role in the invasion of potentially invasive non-native species. Byers et al. (2002) propose that characterizing the factors limiting the persistence of a species in its area of introduction, and particularly on the invasion front, can simulate the conditions determining its geographical distribution.

Dukes & Mooney (1999) highlight the importance of climate change in increasing biological invasions. With its direct and indirect effects on ecosystems, allochthonous and autochthonous species are reacting in a variety of ways. Some are favored and expand their populations to the detriment of others.

I.6 Impacts of invasive species

The ecological, economic and health impacts of invasive species, particularly invasive plants, can be difficult to determine (Bifolchi, 2007) and their magnitude varies according to the invasive species and/or ecosystems invaded. Depending on the characteristics possessed by a plant, its invasion leads to varying degrees of damage to biodiversity, and can result in the local decline or even extinction of native species or the modification of habitats (UICN/PACO, 2013).

The ecological and even economic value of ecosystems depends on the existence of highly diversified species communities (Jacques et al., 2006). However, one of the major impacts of biological invasions is the homogenization of ecosystems by reducing the specific richness of species development stages (Habonayo et al., 2019). This in turn reduces the essential ecosystem services of an invaded ecosystem.

When it comes to invasive alien species, the damage is much greater, as summarized by Soubeyran (2008). They can include :

- alteration of the functioning of the natural or semi-natural ecosystem at the level of ecological processes;
- at the level of ecosystem composition, the regression, hybridization or disappearance of native species;
- in terms of economic activities, penalize agricultural yields or the tourist value of landscapes;
- in terms of human health, cause allergies or promote the transmission of pathogens (viruses, bacteria).

I.7. Biological invasions in Burundi

The proliferation of invasive species is a reality in Burundi, where they have a major impact on the diversity of natural ecosystems and agro-ecosystems. The causes identified for this proliferation are mainly uncontrolled introductions of exotic species, the rapid replacement of agricultural breeds and varieties in use, ecological imbalance due to overexploitation of biological resources and climate change (Masabo & Nindorera, 2019).

Indeed, the introduction of exotic species into certain protected areas is currently a real threat. For example, the proliferation of water hyacinth in Lake Tanganyika and Lake Rweru in the Northern Protected Landscape has become increasingly worrying. The invasion of Rusizi National Park and Murehe Natural Forest by *Lantana camara* has already led to the suppression of several types of plant formations.

The expansion of the invasive liana *Sericostachys* scandens (Habonayo et al., 2019) in Kibira National Park and Bururi Forest Nature Reserve, where it has a strong influence on the diversity of forest stands, is an eloquent example of the effects of ecological imbalance (Zihalirwa et al., 2020). This imbalance favours the proliferation of certain species (even indigenous ones) to the detriment of others, due to the absence of a regulator (eliminated competitor or predator).

CHAPTER II: MATERIALS AND METHODS

II.1 Introduction to the study area

The Ruvubu River is part of the greater Nile River basin, as Burundi's hydrological network comprises two major hydrological basins:

- the Congo Basin, the world's second largest river after the Amazon;
- the Nile basin, the continent's longest river, with its southern source in Burundi (MEEATU, 2014).

Most (over a quarter) of Burundi's water supply to the Nile is drained by the Ruvubu, Burundi's most important watercourse and the river's southernmost tributary (Bizuru, 2005).

Part of the Upper Akagera basin, the Ruvubu river has its source in the mountains of the Congo-Nile ridge. Along its 285 km length in Burundi, the Ruvubu receives numerous tributaries, the main ones being the Kinyankuru, Ndurumu, Nyakigezi, Mubarazi, Ruvyrironza, Nyabaha and Kayongozi. Most of these rivers have their source on the Congo-Nile ridge. The Ruvubu passes through three ecologically different natural regions: Buyenzi, Kirimiro and Bweru. (APRN/BEPB, 2012).

The Ruvubu River flows through a national park bearing its name, and is listed as one of Burundi's four Ramsar sites. The Ruvubu is a meandering river, with little flow between the entrance and exit of the park. The watershed of the Ruvubu represents 10,200 km^2 and the Ruvubu marshes are largely flooded and occupied by permanent swamps (Bizuru, 2005; Masharabu, 2011). Figure II.1 shows the location of the Ruvubu River and our study area.

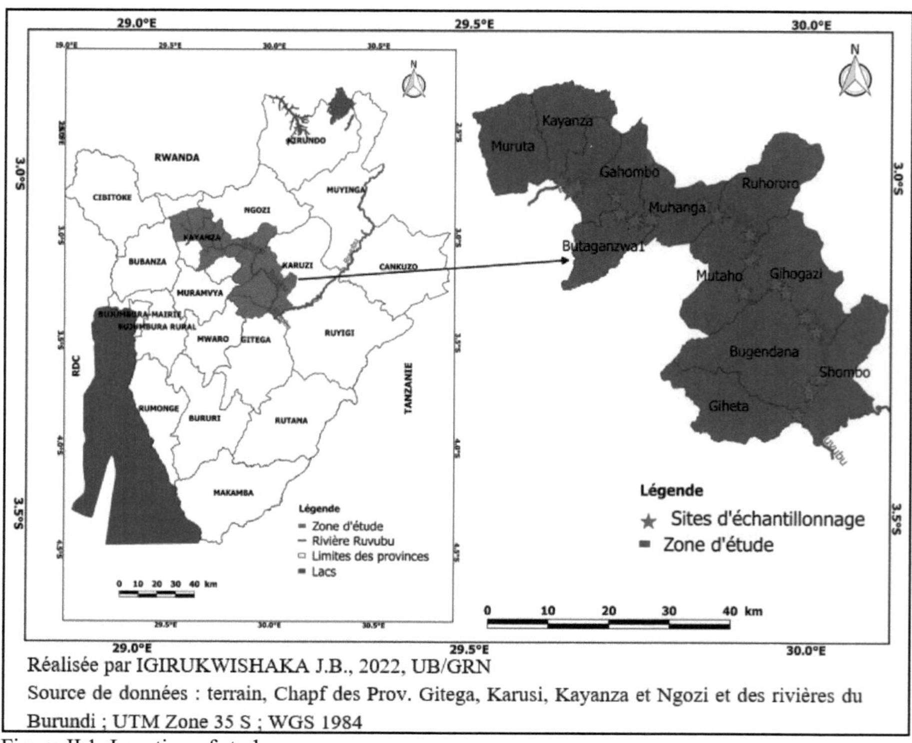

Réalisée par IGIRUKWISHAKA J.B., 2022, UB/GRN
Source de données : terrain, Chapf des Prov. Gitega, Karusi, Kayanza et Ngozi et des rivières du Burundi ; UTM Zone 35 S ; WGS 1984

Figure II.1. Location of study area

II.2. Data collection equipment

The material used in this study consists of :

- Equipment for the flora inventory: decameter (50 m), newspaper and mechanical press, pruning shears, a register to record observations;
- A Garmin GPS receiver to record geographic coordinates;
- Excel 2013, MVSP (Multi Variate Statistical Package) and QGIS 3.4.12 software used for data processing, analysis and cartographic representation of the study area.

II.3. Data collection

The phytosociological approach to data collection was favoured. This is because, according to Bouzillé (2007)in the current context of concern for biodiversity, phytosociological data can be used as a reference for diagnosing the state of biodiversity of natural habitats.

Sampling was carried out at semi-natural sites in the marshes of the Ruvubu River. The sampling area stretched from the bridge of the national road n°1 (RN1) to the bridge of the national road n°12 (RN12), all on the Ruvubu river (Fig.II.1). The choice of such sites was motivated by their high level of disturbance due to their accessibility, facilitated by the presence of bridges (Rejmánek et al., 2013).. Data were collected from 38 phytosociological surveys in the vicinity of bridges crossing the river, divided into 19 sites.

In each of these locations, an area was selected as representative for harvesting, based on the floristic and physiognomic homogeneity of the vegetation, i.e. the structure and composition of the vegetation, reflecting ecological homogeneity. Harvesting was carried out using the minimum area technique, and survey boundaries were set to avoid areas of contact between different plant species.

II.3.1. Determining the minimum area

The minimum area designates the surface to be surveyed (on which the plant species present are inventoried) and which is representative, i.e. containing almost all the species in the plant community, and beyond which the number of species no longer increases.

According to Meddour (2011)the minimum area in a herbaceous environment must be less than 100 m^2 . In the present study, we considered a minimum area varying between 32 m^2 and 64 m^2 depending on the diversity of species present on an area to be surveyed and their degree of homogeneity.

To determine the minimum area, the classic method was used. This consists of delimiting an area of 1 m^2 in a carpet of plant cover where homogeneity has been recognized, and recording all the species growing there. The area is then doubled each time, taking care to record all the species that appear until the area at which the number of species no longer varies.

The principle is to consider the number of species as a function of surface area. The list of species recorded increases as the survey area increases, but in a non-linear way. After a certain extent, the curve representing the number of species as a function of the area surveyed shows a plateau. The beginning of this plateau corresponds to the minimum area where we can be sure, with a high probability, of having observed almost all the species in the group.

In each survey, information on floristic data is collected. This includes the list of species present and, for each species, an estimate of its cover according to the Braun-blanquet (1932).

II.3.2 Sample identification

In the field, specimens of all species were collected and placed in herbariums. The scientific names of the species were determined as a preliminary step, drawing on the knowledge of the guides (three older people chosen from each locality) who provided the vernacular names. We then consulted the works of Merlier & Montegut (1982); Reekmans & Niyongere (1983); Troupin (1978, 1983, 1985, 1988) ; LeBorgeois & Merlier (1995); Fischer & Killmann (2008) and Habiyaremye & Nzigidahera (2016); Nzigidahera et al. (2020). In addition, the samples are compared with the collections of the herbarium of the Faculty of Sciences at the University of Burundi.

II.4. Data processing

II.4.1. Floristic data

To check the homogeneity of the groupings, a frequency analysis was carried out. A histogram of frequency classes was drawn up. According to Raunkiaer (1934)you can tell whether a table is homogeneous or not by the shape of the histogram. If the histogram is in the shape of an inverted L, U or J, then the table can be considered homogeneous, otherwise it is heterogeneous.

Analysis of the flora began by establishing the species composition of the environment (specific richness) and calculating diversity indices: Shannon's diversity (H'), Simpson's diversity (D) and Piélou's regularity (equitability) (E).

Shannon index (H')

Shannon's diversity index is an index derived from information theory that takes into account relative species abundance and total species richness (Ricklefs & Miller, 2005). Taking species abundance into account when calculating the diversity index is essential, as their functional roles in the community vary according to their abundance. This index is chosen because, according to Ramade (2009)it is relatively independent of sample size. It is given by the expression :

$$H = \frac{\Sigma(ni)}{N} \log_2 \frac{ni}{N} \quad \text{or} \quad H = -\sum pi \log_2 pi$$

where $\begin{cases} n_i : \text{species abundance} \\ N: \text{total number of species} \\ p_i : \text{the proportional abundance of each species} \end{cases}$

The Shannon index is zero when there is only one species, and its maximum value is equal to $\log_2 N$ when all species have the same abundance. (Dajoz, 2006).

Simpson Index (D)

Simpson's index is a dominance index. On the one hand, the value 1 is reached if there is only one species present (N=1), i.e. there is complete dominance. On the other hand, values tending towards 0 are obtained if there is a large number of species (absence of dominance). This index is given by the ratio :

$D = \frac{\sum(ni)^2}{N}$ or $D = \sum(pi)^2$ where

- N: the set of all values of defined importance
- n_i : the importance value of each constituent of each species
- p_i : probability of each part of the set, or relative abundance of each species

The diversity of this Simpson index is given by its reciprocal index (1-D), so that a high index reflects high diversity.

Equitability (E) or Piélou index

The equitability or regularity index reflects species stability. It measures the degree of diversity achieved by a stand in relation to its maximum value, and can be used to compare two groups that do not have the same number of species. It also uses, in a non-direct way, the ratio n_i /N and its maximum value is equal to 1. It is obtained by :

$$E = \frac{H}{\log_2 N}$$

Using these three indices concomitantly enables a more comprehensive study of information concerning community structure (Grall & Coic, 2006).

Non-taxonomic descriptors (biological types, diaspore types, phytogeographical types) were also used to analyze the flora.

For biological types (TB), the system of Raunkiaer (1934) as modified by Lebrun (1947) was used. This system is used by other authors (Bangirinama, 2010; Bizuru, 2005; Hakizimana et al., 2012; Masharabu, 2011; Nduwimana, 2014) and recognizes Phanerophytes (Ph),

Chamephytes (Ch), Hemichryptophytes (Hc), Therophytes (Th), Geophytes (Ge) and Hydrophytes (Hy).

For phytogeographical types, the system of Lebrun (1947) modified by (White, 1979, 1983) was used. This system has been used in work carried out in Burundi (Bangirinama, 2010; Bizuru, 2005; Hakizimana et al., 2012; Masharabu, 2011; Nduwimana, 2014) where we recognize species with a wide distribution across the globe (Cos), pantropical species (Pan); paleo-tropical species (Pal); afrotropical species (Afr trop); African multi-regional species (Plur Afr); montane species (Mo), link species (Li Mo-SZ, Li G-Mo); sudanozambezian species (SZ).

For diaspore types, sexual reproduction has been considered. The diaspores concerned are fruits or seeds. The classification system used is that of Dansereau & Lems (1957) which recognizes Sarcochores (sarco), Desmochores (Desmo), Sclerochores (Sclero), Pterochores (Ptero), Pogonochores (Pogo), Ballochores (Ballochores), Barochores (Baro) and Hydrochores (Hy).

For these non-taxonomic descriptors, weighted spectra were established. The weighted spectrum (WS) in % is the ratio of the sum of the overlap values (Ri) of all species (U) presenting the life trait to the sum of the overlap values of all species (N).

$$SP = \frac{\sum_{i=1}^{u} R_i}{\sum_{i}^{N} R_i} . 100$$

The MVSP (Multi Variate Statistical Package) software was used to individualize the groupings. The latter segregates surveys on the basis of their affinity to each other, and characterizes the groupings formed by species and their abundance-dominance index.

II.4.2. Identification of invasive species

Determining which species were invasive required a combination of 3 criteria: average cover rate, species frequency and the biological characteristics of the invasive specific plasticity of the dominant species.

II.4.2.1. Recovery rates

The first indicator of an invasive species in an environment is its high degree of dominance. In this study, the average space (cover) covered by all species in all surveys was calculated. The overlap rate expresses the transformation of species abundance-dominance coefficients into a semi-quantitative value for each species in its environment. It consists in finding the average space (cover) covered by all the species (Emj) in all the surveys; it is given by the following expression :

$$Emj = \frac{\sum_{ni}^{N} RMi}{P}$$

where N= total number of species
RMi: average cover of each species
P: total number of readings

According to th[e] ... as a cover rate of 75-100%; in this study, the average cover for a dominant species is given by the formula Emj X 75%, according to the lower bound of abundance-dominance, assuming that this space is dominated by a single species. For the purposes of this study, an invasive species is any species whose cover is equal to or greater than 2%. (Blanfort et al., 2009; Van Kleunen et al., 2010).

II.4.2.1. Frequency

The frequency of listed species was analyzed. Thus, a species with a frequency equal to or greater than 60% is considered invasive (Blanfort et al., 2009; Van Kleunen et al., 2010).

II.4.2.2. Biological characteristics of specific plasticity

The characterization of a biological invasion must refer to a certain number of criteria. The characteristics taken into account in this work are grouped into eight categories. They are : 1° sexual reproduction (by entomogamous or bisexual flower); 2° asexual or vegetative reproduction (by leaf, cutting, rhizome, stolon); 3° dissemination (anemochory, zoochory, hydrochory,...); 4° seed bank (fruits per stem ≥ 10, seeds per pod ≥ 10); 5° seed size (> or < eleusine size); 6° weather resistance (presence of spines or hairs, water reserve, fire resistance, bitter taste); 7° lifespan (annual or multiannual); 8° absence or presence of predators (Van Kleunen et al., 2010).

The analysis of the biological characteristics of invasive specific plasticity concerned presumed invasive species according to their average cover rate (2%) and the frequency (60%) of dominant species.

A species will have the maximum chance of being invasive when all the characteristics of these categories are present. It will have the least chance if one of the characteristics in each category is present, but it should be noted that a characteristic absent in one category can be replaced by a characteristic from another category. In this case, the species can have a number of characteristics greater than or equal to 8, regardless of the characteristic categories.

a. Reproduction

Plants can reproduce sexually, asexually or both. Sexual reproduction is ensured by sexual organs located in the flower. In invasive plant species, high reproductive capacity is an important characteristic. Asexual reproduction also plays a major role in invasive species. Asexual reproduction can take the form of cuttings, rhizomes, stolons or even leaves. The combination of the two pathways can account for a more dynamic demography and faster propagation capacity in invasive plant species.

b. Dissemination

There are several recognized dispersal vectors: gravity, wind, water, animals or expulsion by the plant itself. The study considered wind, water and animals as long-distance dissemination vectors.

c. Seed bank and size

A plant producing a large number of seeds (diaspores) has a good chance of rapidly colonizing the environment. The ease of dissemination depends on the size of the seeds (diaspores). The reference used in this study is the size of an eleusine seed (about 1mm in diameter). Wind easily disperses seeds smaller than or equal to the reference size (Bavumiragiye & Niyonkuru, 2018)..

To understand the reproductive and dissemination adaptations of the species inventoried, our work has drawn on studies by other authors (Muoghalu & Chuba, 2005; Agboola & Joseph, 2014 and Lisan, 2014) on seed bank, seed size and weather resistance.

d. Weather resistance

Invasive species display a number of adaptations to cope with changing ecological conditions, but also to ward off predators. In this study, the types of adaptations taken into consideration are the presence of spines or hairs, water reserve, fire resistance, bitter taste, repulsive odor.

e. Service life

The lifespan of a species influences flowering time. If a plant's lifespan is long, it's more likely to have several flowering periods and produce many diaspores. Or, if the lifespan is short, the plant will mature early and produce diaspores very early. Thus, we have seasonal, annual or multi-annual plants.

f. Absence of enemies (predators, pathogens, etc.)

Equilibrium in nature results from interactions between the components of a biocenosis, but also between the same biocenosis and its biotope. This balance can be disrupted by the introduction of a foreign element or the elimination of a biocenosis component.

In fact, an invasive plant originating from a distant territory (geographically and/or ecologically) has a low probability of encountering enemies (predators, parasites, etc.) in its new territory, which would have limited its growth in its place of origin.

Furthermore, the elimination of a predator of a species in an ecosystem can allow the proliferation of populations of this prey species, which in turn can have harmful ecological consequences. Having identified invasive species using the above criteria, to validate their invasive nature, we made particular use of the document on the status of invasive species in Burundi by (Nzigidahera, 2017)the International Union for Conservation of Nature's Global Invasive Species Database[1] :(IUCN-GISD, 2021; IUCN, 2021).

[1] https://www.iucn.org/content/global-envahissante-species-database-gisd and http://www.iucngisd.org/gisd/

CHAPTER III: PRESENTATION OF RESULTS

III.1. Sampling effort

The list of species recorded increases as the survey area increases, but in a non-linear way. From 32 m^2 , the curve representing the number of species as a function of the area surveyed shows a plateau, the minimum area (Fig.III.1).

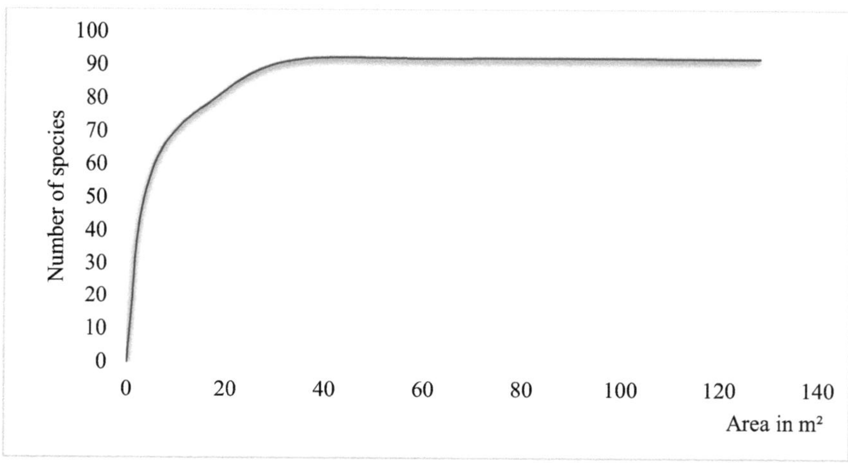

Figure III.1. Area-species curve sampling effort

Analysis of the frequency classes shows that sampling was carried out in a homogeneous environment, as illustrated by the inverted J shape of the histogram (Fig.III.2). This analysis shows that 21 species are in class V (frequency of at least 80%), 19 in class IV (60 to 79% frequency), 27 in class III (40 to 59% frequency), 42 in class II (20 to 39% frequency) and 60 in class I (frequency less than 20%).

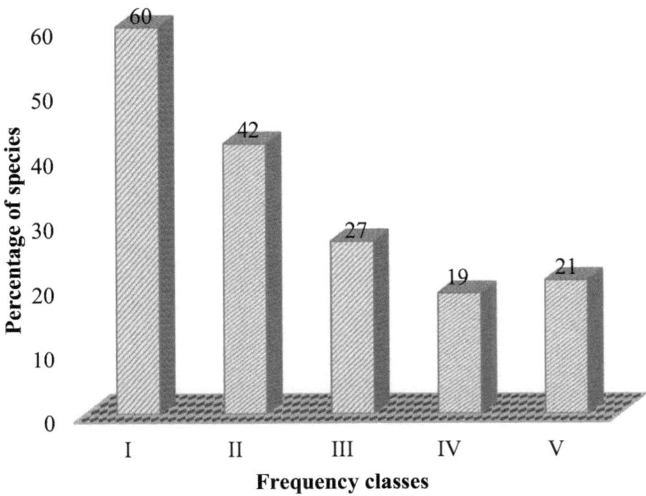

Figure III.2. Histogram of frequency classes

III.2 Floristic inventory

III.2.1. Floristic summary

The floristic report shows 161 plants in 54 families and 125 genera (appendix 2) at 38 sites (appendix 1). Dicotyledons are best represented (70.18%). Monocotyledons accounted for 26.70%, while Pteridophytes were poorly represented (3.10%) (Table III.1).

Table III.1. Distribution of plants inventoried in the Ruvubu river marshes in higher taxa

		Workforce		
Branch	**Classes**	**Families**	**Genres**	**Species**
Pteridophyta	**Filicopsida**	4 (7,40%)	4 (3,2%)	5 (3,10%)
Magnoliophyta	**Liliopsida**	8 (14,81%)	28 (22,4%)	43 (26,70%)
	Magnoliopsida	42 (77,77%)	93 (74,4%)	113 (70,18%)
Totals		**54 (100%)**	**125 (100%)**	**161 (100%)**

The families with the highest number of species are, in descending order, Poaceae with 29 species (17.39%), Asteraceae with 24 species (14.91%), Fabaceae with 17 species (10.56%), Cyperaceae with 8 species (4.97%), Malvaceae with 6 species (3.73%) and Euphorbiaceae with 5 species (3.11%) (Fig. III.3).

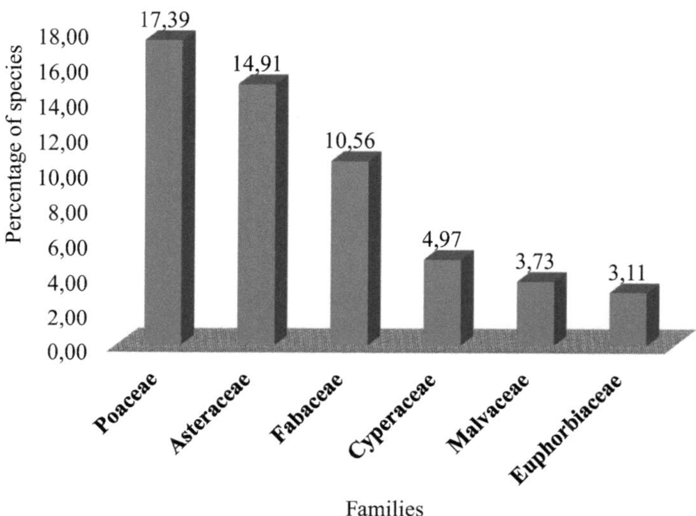

Figure III.3. Specific richness of the most represented families in the Ruvubu marshes

III.2.2. Biological types

The results of calculating the weighted spectrum of biological types place champhytes in first place with 33.73% of all biological types, followed by therophytes (19.70%) and geophytes (16.93%), then phanerophytes (15.78%) and hemicryptophytes (13.77%). Hydrophytes come last (0.096%) (Fig.III.4).

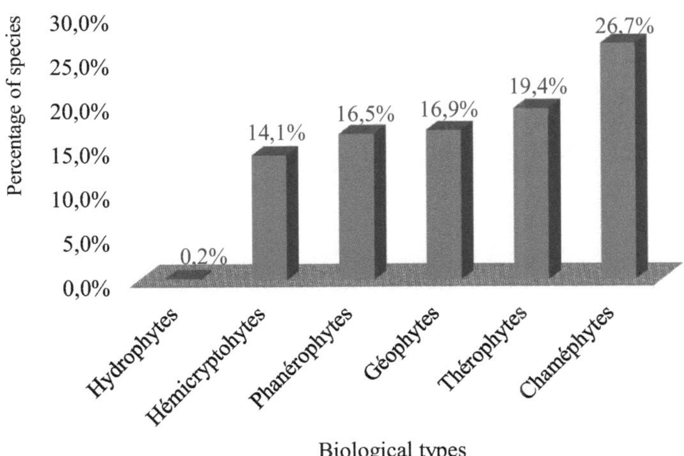

Figure III. 4. Weighted spectrum of biological types

III.2.3. Types of diaspora

Analysis of diaspore types (Fig.III.5) showed that anemochores (pogonochores, pterochores and sclerochores) accounted for 56.37%, autochores (ballochores and barochores) for 23.49% and zoochores (sarcochores and desmochores) for 20%.

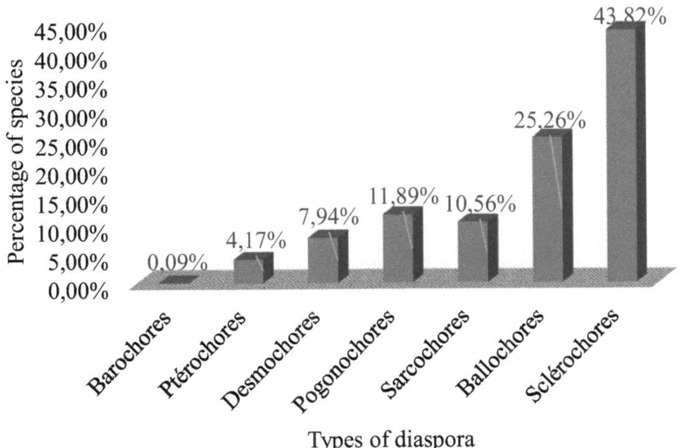

Figure III. 5. Weighted spectrum of diaspore types

III.2.4. Phytogeographical types

Observation of phytogeographical types (Fig.III.6) reveals that widely distributed species (Cos, Pan, Pal, Subcos, Afr-Am) are the most abundant at 62.04%. They are followed by African multi-regional species (Plur Afr, Afro-Trop, Afr-Mal) with 19.42% and regionally distributed species (Mont, SG, SZ) with 11.70%). Next come introduced American species (5.30%) and finally link species (L.SZ-G, LSZ-Mo) with 1.54% of species inventoried.

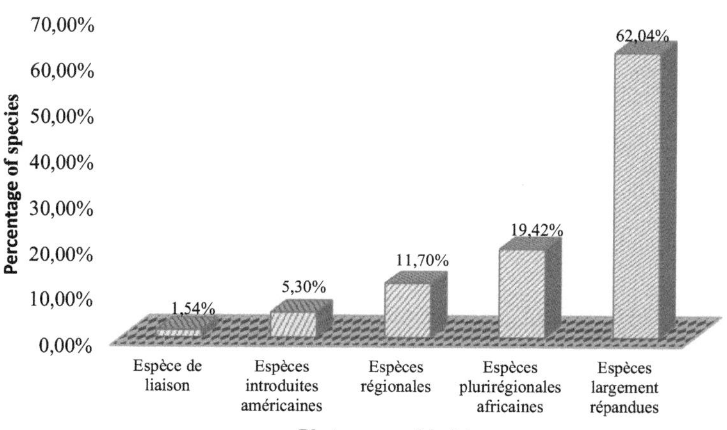

Figure III. 6. Weighted spectrum of phytogeographical types

III.2.5. Diversity indices

The results of calculating the Shannon (H), Simpson (D) and Equitability (E) diversity indices are shown below.

Table III.2. Values of observed species richness and floristic diversity indices found in the Ruvubu River marshes

Diversity indexes	H'	1-D	E	N
value	5,946	0,982	0,990	161

This table shows that the Shannon diversity index and the Equitabity index are relatively high, being more than half their maximum value of 3.66 and 0.5 respectively. The maximum value of the Shannon diversity index is $\log_2 N = 7.33$. As for Simpson's diversity index, its value (0.96) tends towards the maximum value, which is equal to 1.

III.2.6 Group individualization

Hierarchical ascending classification using the Bray-Curtis dissimilarity index shows a separation of four groupings per plant community (Fig.III.7). The dissimilarity of these groupings is between 30% and 40%.

Reconstitution of the four groups shows that the group with *Ageratum conyzoides* L. and *Oplismenus burmanni* (Retz.) P.Beauv. (G1), comprising 13 records, has 83 species, with *Carissa spinarum* L., *Acanthus polystachyus* Del., *Combretum collinum* Fresen., *Eragrostis tenuifolia* (A.Rich.) Hochst. ex Steud., *Oplismenus burmanni* (Retz.) P.Beauv., *Ageratum conyzoides* L.

The grouping with *Centella asiatica* (L.) Urb. and *Galinsonga parviflora* Cav. (G2), made up of 9 records, has 67 species with a high frequency of *Ageratum conyzoides* L., *Centella asiatica* (L.) Urb, *Commelina Africana* L., *Aspilia pluriseta* Schweinf., *Asplenium onopteris* L., *Commelina benghalensis* L., *Galinsonga parviflora* Cav, *Mimosa pigra* L., *Mimosa diplotricha* C.Wright, *Hyparrhenia dichroa* (Steud.) Stapf.

The grouping with *Centella asiatica* (L.) Urb. and *Cynodon dactylon* (L.) Pers. (G3), consisting of 6 surveys, has 77 species, with *Bidens pilosa* L., *Centella asiatica* (L.) Urb., *Commelina Africana* L., *Aspilia pluriseta* Schweinf..., *Asplenium onopteris L., Commelina benghalensis L., Galinsonga parviflora Cav, Asplenium onopteris* L., *Commelina benghalensis* L., *Galinsonga parviflora* Cav., *Cynodon nemfluensis* Vanderyst, *Cynodon dactylon* (L.) Pers, *Gynandropsis gynandra (*L.) Briq, *Polygonum glabrum* Willd., *Panicum maximum* Jacq, *Nephrolepis undulata* (Afzel. ex Sw.) J. Sm.

And the grouping with *Pennisetum trachyphyllum* Pilg. and *Nephrolepis undulata* (Afzel. ex Sw.) J. Sm. (G4) has 10 records with 92 species and a high frequency of *Bidens pilosa* L., *Centella asiatica* (L.) Urb., *Commelina Africana* L., *Commelina benghalensis* L., *Galinsonga parviflora* Ruiz & Pav.., *Cynodon nemfluensis* Vanderyst, *Cynodon dactylon* (L.) Pers, *Polygonum glabrum* Willd, *Panicum maximum* Jacq, *Nephrolepis undulata* (Afzel. ex Sw.) J. Sm, *Imperata cylindrica* (L.) Beauv, *Kyllinga erecta* Schumach, *Pennisetum trachyphyllum* Pilg, *Lantana camara* L., *Oxalis corniculata* L.

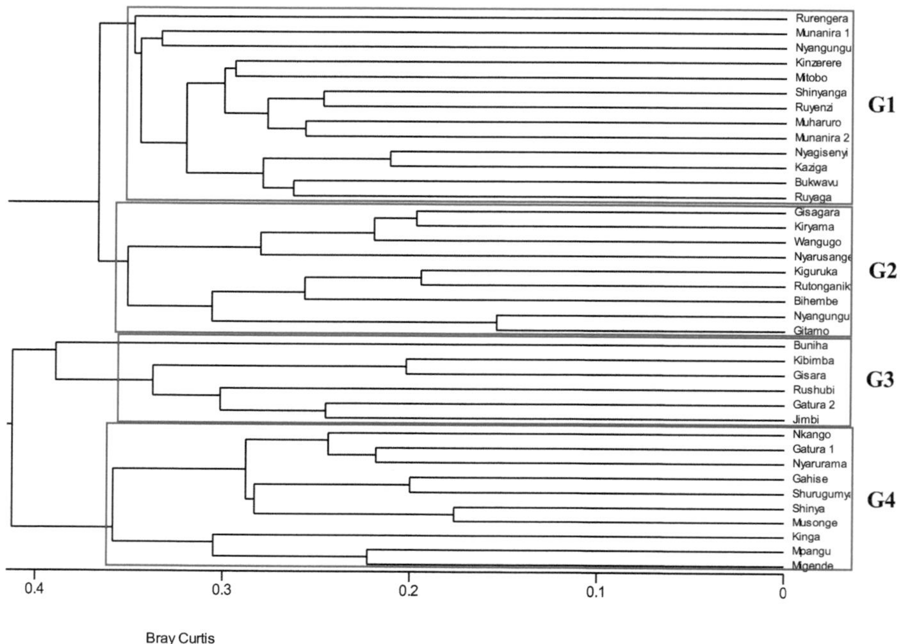

Figure III. 7. Group hierarchy

III. 3 Invasive plant species determined

III.3.1. Coverage rate of the most common species

The first indicator of invasiveness is a species' high degree of cover and frequency in an environment. For any species with a cover rate greater than or equal to 1, its frequency is tested (Fig.III.8).

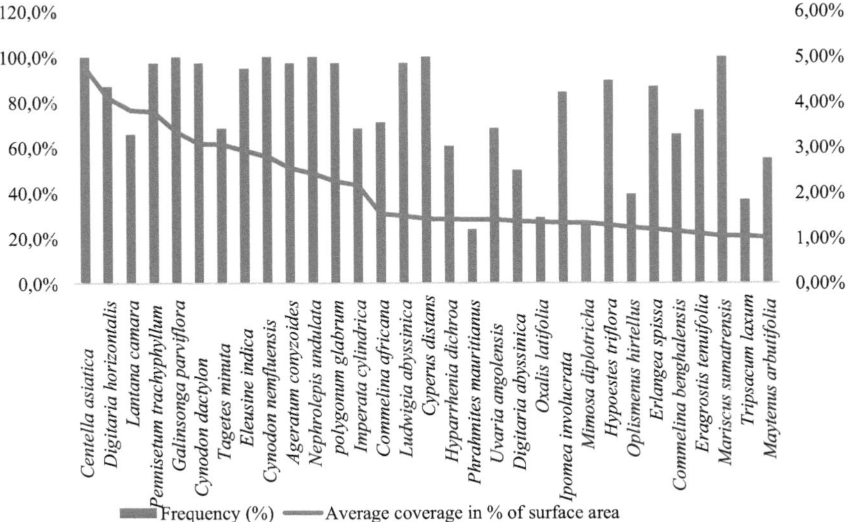

Figure III. 8. Average cover rate and frequency of suspected invasive species

The combination of average cover rate (2%) and frequency (60%) of dominant species identifies 13 invasive species. These are the species listed in Table III.3, with their respective mean cover rates and frequencies.

Table III. 3. Presumed invasive species

Species	Recovery	Frequency
Centella asiatica (L.) Urb.	4,79%	100
Digitaria horizontalis Willd.	4,10%	85
Lantana camara L.	3,82%	67,5
Pennisetum trachyphyllum Pilg.	3,79%	97,5
Galinsonga parviflora Cav.	3,35%	100
Cynodon dactylon (L.) Pers.	3,09%	97,5
Tagetes minuta L.	3,07%	65
Eleusine indica (L.) Gaertn.	2,93%	92,5
Cynodon nemfluensis Vanderyst .	2,80%	97,5
Ageratum conyzoides L	2,55%	92,5
Nephrolepis undulata (Afzel. ex Sw.) J. Sm.	2,43%	100
Polygonum glabrum Willd.	2,26%	97,5
Imperata cylindrica (L.) Beauv.	2,17%	70

III.3.2. Biological characteristics of the invasive specific plasticity of dominant species

a. Reproduction

Presumed invasive species reproduce both asexually (vegetatively) and sexually, with the exception of *Tagetes minuta*, *Eleusine indica* and *Ageratum conyzoides*, which reproduce sexually only; and *Nephrolepis undulata*, which reproduces asexually.

b. Dissemination

Wind, water and animals are vectors for long-distance seed dissemination. Analysis of seed dispersal patterns of suspected invasive species shows that more than half are wind-borne (anemochores: 50% sclerochores and 10.71% pogonophores). Water and animals play a small part in the spread of these plants. Indeed, only *Lantana camara* is dispersed by birds (endozoochory and dyszoochory) and *Centella asiatica* by water (hydrochory).

c. Number and size of seeds

Observation of the seeds and their size shows that, with the exception of *Lantana camara,* which has large fruits and seeds compared with the reference size (eleusine seed size), all the other presumed invasive species have seeds around the reference size. In terms of numbers, the seeds can be counted in the hundreds per stem.

d. Weather resistance

The species presumed to be invasive in this study do not show any particular adaptations to bad weather in morphological terms, but *Lantana camara* shows a more pronounced adaptation. *Lantana camara* has thorns, and its seeds are protected by a thicker, harder coating. Its stems are more resistant to drought and retain their germinative power for longer.

e. Service life

Presumed invasive species are seasonal plants, with the exception of *Lantana camara*, which is annual or even multi-annual.

f. Absence of enemies (predators, pathogens, etc.)

The species *Digitaria horizontalis, Pennisetum trachyphyllum, Cynodon nemfluensis*, *Cynodon dactylon and Eleusine indica* can be used as fodder by ruminant animals, while *Galinsonga parviflora* can be used by rodents. Other species have developed defense mechanisms against all forms of predation. These include the emission of repellent odors (*Lantana camara, Tagetes minuta, Ageratum conyzoides*, and *Polygonum glabrum*) and/or thorns (*Lantana camara* and *Imperata cylindrica*).

The combination of average cover rate, frequency and biological characteristics of invasive specific plasticity shows that 13 species can be qualified as invasive species in the marshes of the Ruvubu river, which is the study area. These are *Centella asiatica, Digitaria horizontalis, Lantana camara, Pennisetum trachyphyllum, Galinsonga parviflora, Cynodon dactylon, Tagetes minuta, Eleusine indica, Cynodon nemfluensis, Ageratum conyzoides, Nephrolepis undulata, Polygonum glabrum* and *Imperata cylindrica.*

In addition, some species show signs of invasiveness if the criteria are considered separately. Taking into account the average cover rate, *Oxalis latifolia, Mimosa diplotricha, Mimosa pigra, Xanthium strumarium, Panicum maximum, Acanthospermum australe, Kyllinga erecta, Phragmites mauritianus* and *Paspalum notatum* appear to be the most overgrown in their respective environments, although they are not common throughout the study area. However, among these, 2 species (*Kyllinga erecta* and *Phragmites mauritianus*) are rejected as characteristic of marshes or flooded areas.

III.3.3. Characterization of selected invasive plants

The species selected as invasive are the following: *Centella asiatica, Digitaria horizontalis, Lantana camara, Pennisetum trachyphyllum, Galinsonga parviflora, Cynodon dactylon, Tagetes minuta, Eleusine indica, Cynodon nemfluensis, Ageratum conyzoides, Nephrolepis undulata, Polygonum glabrum* and *Imperata cylindrica* (Fig.III.9).

Oxalis latifolia, Mimosa diplotricha, Mimosa pigra, Xanthium strumarium, Panicum maximum, Acanthospermum australe and *Paspalum notatum* are invasive species in their respective environments on the surveyed site. They are potentially invasive in the area (Fig.III.10).

Figure III. 9. Images of the 13 invasive species selected in the Ruvubu marshes

Figure III. 10. Images of the 7 potentially invasive plants in the marshes of the Ruvubu

III.3.3.1 Floristic gradient of selected invasive plants according to survey sites

By analyzing the floristic gradient of invasive plants selected according to survey sites, the trend of the curve shows that the number of invasive plants increases from upstream to downstream (Fig.III.11).

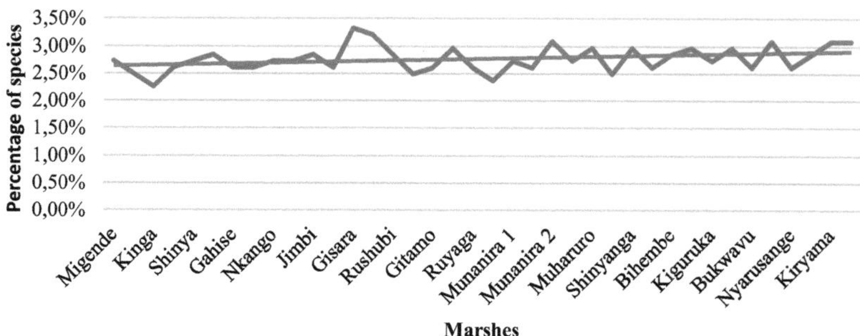

Figure III.11. Floristic gradient of selected invasive species by survey site

III.3.3.2. Representative families of invasive species

Analysis of the families representing invasive species shows that these species fall into six families (Poaceae, asteraceae, Apiaceae, Nephrolepidaceae, Polygonaceae and Verbenaceae), with Poaceae (46.15%) and asteraceae (23.08%) predominating over the other families with 7.69% each (Fig. III.12).

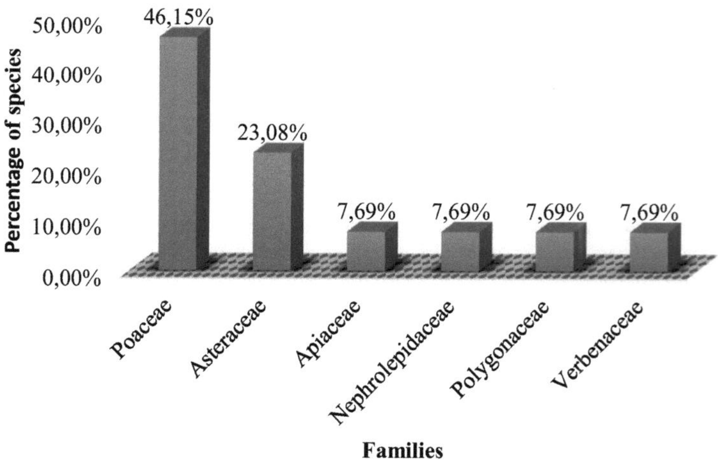

Figure III. 12. Species richness of the most common families of invasive species

III.3.3.3. Phytogeographical types of invasive species

All thirteen selected invasive species are divided into paleo-tropical species (70%), afro-tropical species (23.33%), montane species and American species with 3.33% each (Fig.III.13).

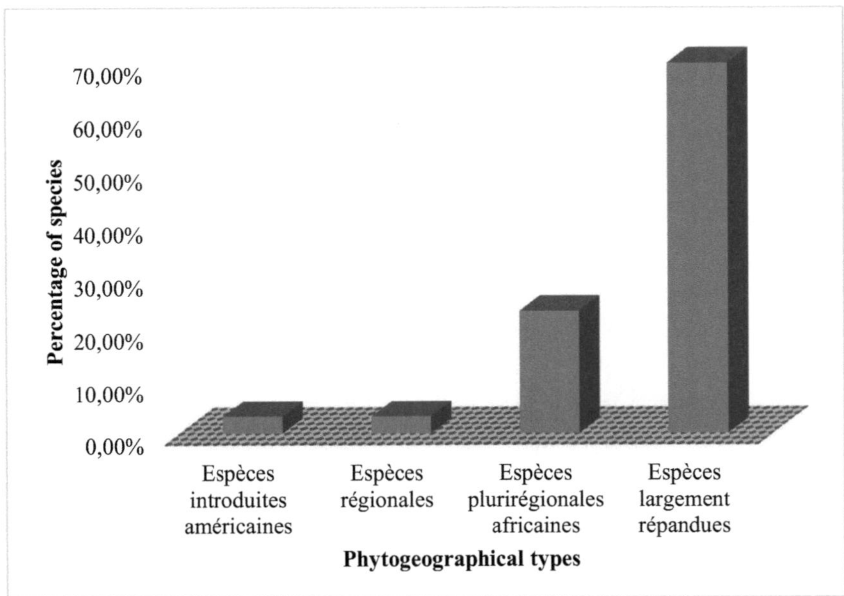

Figure III. 13. Rough spectrum of phytogeographical types of selected invasive species

III.3.3.4. Biological types of selected invasive species

The results of the calculation of the gross spectrum of biological types of invasive species place champhytes in first position with 38.71% of all biological types, followed by geophytes and hemicryptophytes (19.35%) for each type. Therophytes (12.90%) and phanerophytes (9.68%) come fourth and fifth respectively (fig.III.14).

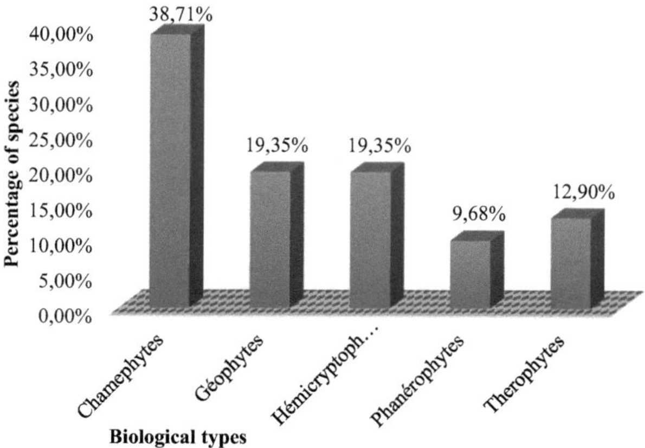

Figure III. 14. Crude spectrum of biological types of selected invasive species

III.3.3.4. Types of diaspores of invasive species

In terms of sexual propagation, most of the invasive species selected are spread by the wind at over 60% (sclerochores: 50.00% and pogonochores: 10.71%), by the plant itself (ballochores) at 28.57% or by animals at 10% (sarcochores: 7.14% and desmochores: 3.57%) (Fig.III.15).

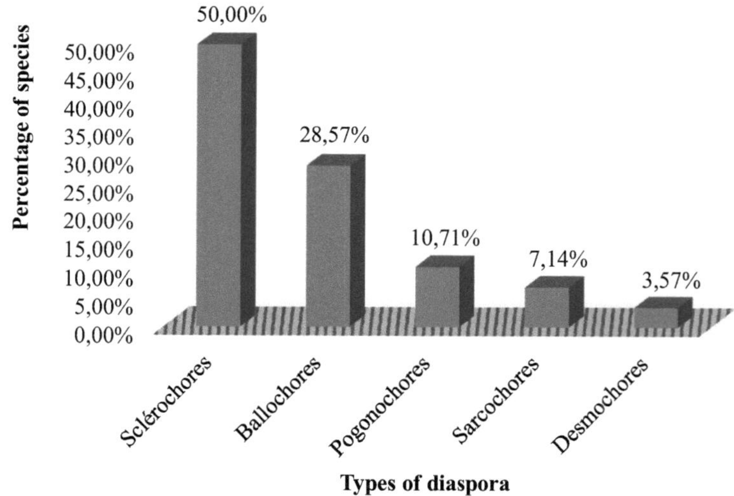

Figure III. 15. Crude spectrum of diaspore types for selected invasive species

CHAPTER IV: DISCUSSION OF RESULTS

IV.1. Floristic composition

The results show that the Magnoliopsida class dominates the taxa of the Ruvubu River marshes, with 113 species in 93 genera and 42 families. Meanwhile, the Liliopsida class (monocotyledons) is represented by 43 species in 28 genera and 8 families. The Filicopsida class (ferns) is less well represented, with 5 species in 4 genera and families. These results coincide with those of Bigendako's assessment of the vascular flora of Burundi (Bigendako, 1989-1990 in Bizuru, 2005) and other authors who have worked on the flora of Burundi, such as (Dushimirimana et al., 2010; Masharabu et al., 2014; Masharabu, 2011; Nduwimana et al., 2021).

The Poaceae family is most represented in the Ruvubu marshes (17.39%), followed by the Asteraceae (14.91%) and Fabaceae (10.56%) (Fig.4). The predominance of these three families can be explained by their mode of dissemination, with individuals producing a high number of diaspores, and above all their ecological plasticity, enabling them to respond to changes in the environment. These results are similar to those found in the Murembwe river valley by Bavumiragiye & Niyonkuru (2018).

Our results show that the Cyperaceae family is not important in the Ruvubu marshes, although it is more adapted to humid environments (Dushimirimana et al., 2010).. In addition, Rubiaceae, Orchidaceae, Euphorbiaceae and Lamiaceae are less important or absent in the Ruvubu marshes, whereas these families are important in Burundi's flora as a whole (Bizuru, 2005; Dushimirimana et al., 2010). This would indicate a regressive trend in vegetation and a growing anthropogenic footprint in the Ruvubu marshes.

IV.1.1. Biological forms

The results of the weighted spectrum of biological types showed a high abundance of champhytes (33.73%). This chamephytic tendency, characteristic of wet but cold regions, can be explained by the fact that most of the marshes studied are at high altitude. Chamephytes correspond to the stress tolerance strategy, with stress often being trophic, hydric or destructive fire. (Masharabu, 2011). For our purposes, the importance of champhytes is explained by their tolerance to the water factor, given that our study area is flooded during the rainy season.

Therophytes are present at 19.70%. These are annual plants that spend the off-season as seeds or spores. In fact, the marshes of the Ruvubu river are dominated by the two Poaceae and Asteraceae families, which are largely annual plants. The abundance of champhytes and therophytes confirms the herbaceous character of the study area, due to its vegetation degradation and disturbance by repeated cultivation.

It should be noted that the vegetation is also invaded by secondary species, ruderal and cultivated and post-cultivated species as a result of the influence of agricultural activities in the surrounding areas.

IV.1.2. Types of diaspora

Analysis of diaspore types (Fig.III.5) showed that the Ruvubu marshes are dominated by anemochorous plants (pogonochores, pterochores, sclerochores). These results corroborate those of studies carried out in similar environments (Bizuru, 2005; Masharabu, 2011).. In fact, anemochory is a main dissemination strategy for plants in open environments (Nduwimana et al., 2021)such as the marshes of the Ruvubu River.

IV.1.3. Phytogeographical types

In the Ruvubu marshes, widespread species dominate the flora. Analysis of the results by phytogeographical type shows that widespread species account for 62.04%, followed by African multi-regional species (19.42%) and 11.07% of regionally distributed species. This shows that the marshes of the Ruvubu River are being disturbed, which would have led to the disappearance of species from the region as well as colonization by species from other regions. Anthropogenic actions are the driving force behind this disturbance, as concluded by Dushimirimana and colleagues (2010) in the Nyamuswaga marshes.

IV.1.4. Biological diversity

Based on the results of the floristic inventory (Appendix 1), it is remarkable that the marshes of the Ruvubu River show greater species diversity. Indeed, the specific richness of these marshes amounts to 161 species.

Shannon's diversity index is 5.94 and Piélou's regularity index is 0.99. Simpson's diversity index is 0.98. Considering that the potential diversity for 161 species is 7.33 and that the probability of two randomly selected plants belonging to two different species is 0.98, we can conclude that the specific diversity of the Ruvubu River marshes is relatively high. These results corroborate those found by Bizuru (2005) in mountain marshes.

This specific diversity is attributable to the diversity of biotopes existing in the marshes of the Ruvubu River: the marshes studied are located at different altitudes. In addition, the survey sites are located close to bridges and roads crossing the river. As these areas are much frequented and disturbed as a result of their accessibility, they can serve as gateways for new species.

IV.1.5. Grouping hierarchy

Hierarchical ascending classification enabled us to identify four groupings at the site studied. It shows that species are grouped according to altitudinal gradient. In fact, altitude decreases in the direction of flow of the Ruvubu River, and we can see that species are grouped according to their closer specific diversity. The latter, in turn, are formed according to their proximity in altitude.

Group G4 covers surveys at altitudes of between 1,690 m and 1,565 m (in the marshes of the communes of Kayanza, Gatara, Butaganzwa and Gahombo).

Group G3 includes surveys at altitudes of between 1,539 m and 1,525 m (in the marshes of the Muhanga and Ruhororo communes).

Group G1 includes surveys at altitudes between 1503 m and 1461 m (in the marshes of the Ruhororo, Mutaho and Gihogazi communes).

Group G2 includes surveys at altitudes between 1,452 m and 1,411 m (in the marshes of Bugendana, Gihogazi, Shombo and Giheta communes).

However, species richness decreases in the order of these groupings: G4-G3-G1-G2. This finding is similar to that noted by Bizuru (2005) in the marshes of Burundi. Indeed, this author noted that mountain marshes show greater species diversity.

IV.2. Identified invasive species

With regard to the identification of invasive species, the combination of criteria (average cover rate and biological characteristics of invasive specific plasticity) for the dominant species led to the selection of 13 species that could be described as invasive in the marshes of the Ruvubu river, which is the study area. These are *Centella asiatica*, *Digitaria horizontalis*, *Lantana camara*, *Pennisetum trachyphyllum*, *Galinsonga parviflora*, *Cynodon dactylon*, *Tagetes minuta*, *Eleusine indica*, *Cynodon nemfluensis*, *Ageratum conyzoides*, *Nephrolepis undulata*, *Polygonum glabrum* and *Imperata cylindrica*. These results are similar to those of Nzigidahera (2017) who established the situation of invasive species in Burundi. Some of these species are already cited in other localities in Burundi. This is the case of *Cynodon dactylon,* found invasive in the city of Bujumbura by Ndayisaba (2018) and *Eleusine indica* and *Ageratum conyzoides* reported by Bavumiragiye & Niyonkuru (2018) in the Murembwe river valleys. The species *Tagetes minuta*, Ageratum *conyzoides*, *Lantana camara*, *Eleusine indica, Galinsoga parviflora*, etc. are reported even elsewhere as invasive (OSS, 2020)

The dominant invasive plants fall into six botanical families, two of which (Poaceae and Asteraceae) are the most represented, accounting for 69.23% of invasive species. Akodéwou et al (2019)who made a similar observation for these two families, justify their large share in the wealth of invasive species by their high dispersal capacity and great ecological plasticity.

Oxalis latifolia, *Mimosa diplotricha*, *Mimosa pigra*, *Xanthium strumarium*, *Panicum maximum*, *Acanthospermum australe* and *Paspalum notatum* are invasive species in their respective environments on the surveyed site. They have already been cited as invasive species in other regions of Burundi and elsewhere (Osawa et al., 2013; UICN/PACO, 2013; Nzigidahera, 2017; Bavumiragiye & Niyonkuru, 2018; Ben-ghabrit et al., 2018; Akodéwou et al., 2019).. They are potential invasive plants.

Table IV.1 shows some other environments (in Burundi or elsewhere) where these invasive species or potential invaders have already invaded.

Table IV. 1. Some other environments (in Burundi or elsewhere) where these selected or potential invasive species have already invaded

Invasive species	**Some areas already invaded**
Centella asiatica	- Everywhere in Burundi (Nzigidahera, 2017) - In French Polynesia (Fourdrigniez & Meyer, 2008)
Digitaria horizontalis	- In Gambia (CAB International, 2004)

	- In Mayotte (Duperron et al., 2019)
Lantana camara	- Bugesera region (Nzigidahera, 2017) - Martinique, Guadeloupe, Réunion, Mayotte, New Caledonia, Wallis and Futuna, French Polynesia (Soubeyran, 2008) - French Polynesia (Fourdrigniez & Meyer, 2008)
Pennisetum trachyphyllum	- French Polynesia, Australia, California (USA), Colombia, Ecuador, Hawaii, Reunion Island, New Zealand, Peru, Taiwan (Fourdrigniez & Meyer, 2008)
Galinsonga parviflora	- Everywhere in Burundi (Nzigidahera, 2017) - In Brittany (Quéré et al., 2011)
Cynodon dactylon	- Everywhere in Burundi (Nzigidahera, 2017) - In French Polynesia (Fourdrigniez & Meyer, 2008)
Tagetes minuta	- Everywhere in Burundi (Nzigidahera, 2017)
Eleusine indica	- In Tunisia, Algeria, Morocco and Libya (OSS, 2020) - New Caledonia (Vanessa et al., 2009)
Cynodon nemfluensis	- Murembwe Valley, Burundi (Bavumiragiye & Niyonkuru, 2018)
Ageratum conyzoides	- Everywhere in Burundi (Nzigidahera, 2017) - Madagascar (Lisan, 2014) - in Togodo, Togo (Akodéwou et al., 2019)
Nephrolepis undulata	- In Tanzania (Witt et al., 2020) - In Malawi (Mwanyambo et al., 2020)
Polygonum glabrum	- In Normandy, France (Breton, 2014)

Table IV.1: Some other environments (in Burundi or elsewhere) where these invasive species are potential or have already invaded (continued)

Imperata cylindrica	- Everywhere in Burundi (Nzigidahera, 2017) - in Togodo, Togo (Akodéwou et al., 2019)
Acanthospermum australe	- Kumoso depression and Buyogoma and Bweru regions (Nzigidahera, 2017)

Xanthium strumarium	- Rusizi Plain (Nzigidahera, 2017) - In French Polynesia (Fourdrigniez & Meyer, 2008)
Mimosa diplotricha	- Imbo plain, lower Mumirwa escarpments (Nzigidahera, 2017) - Reunion, Mayotte, New Caledonia, Wallis and Futuna, French Polynesia (Soubeyran, 2008) - In French Polynesia (Fourdrigniez & Meyer, 2008)
Mimosa pigra	- Throughout Burundi at altitudes below 2000m (Nzigidahera, 2017) - in Togodo, Togo (Akodéwou et al., 2019)
Oxalis latifolia	- Madagascar (Lisan, 2014) - In French Polynesia (Fourdrigniez & Meyer, 2008)
Panicum maximum	- in Togodo, Togo (Akodéwou et al., 2019) - In French Polynesia (Fourdrigniez & Meyer, 2008)
Paspalum notatum	- In French Guiana (French Guiana Regional Environment Department, 2010)

An analysis of the phytogeographical types of the invasive species selected shows that they are all widespread species, which explains their great colonizing power. What's more, they are all exotic species. Often settling on disturbed land, they can be indicators of this disturbance (Lisan, 2014).

Through their more worrying proliferation (according to the invasive specific plasticity observed) to the detriment of locals, they produce significant changes in composition, structure and consequently, according to Bousquet et al.(2016)ecosystem functioning.

As for the mode of spread, analysis of diaspora types confirms the specific invasive plasticity of the species selected as invasive. These species are characterized by their ability to spread by wind in an open environment such as the marshes of the Ruvubu River, and over long distances (Dushimirimana et al., 2010).. This anemochore mode of spread justifies the great richness of

Poaceae and Asteraceae. The latter are made up of species with small, lighter seeds. In addition, invasive species are adapted to asexual reproduction.

In terms of life forms, champhytes are the most abundant (38.71%) of the invasive species found in the marshes. This form represents an adaptation to withstand the unfavorable conditions of the conquered environment. According to Masharabu (2011)the majority of champhytes correspond to the stress tolerance strategy. In the case of the Ruvubu river marshes, stress is largely hydric, especially with periods of flooding in the rainy season and low water levels in the dry season.

The floristic gradient of invasive plants recorded at the survey sites increases from upstream to downstream. Indeed, the number of invasive species increases with the direction of flow of the Ruvubu River.

On the other hand, species richness increases in the opposite direction to the river flow. Species such as *Aspilia pluriseta, Vernonia lasiopus, Sphaeranthus suaveolens, Setaria pumila, Microglossa pyrifolia, Hypoestes triflora, Gynandropsis gynandra, Guizotia scabra, Euphorbia tirucalli, Emilia caespitosa, Drymaria cordata, Crassocephalum vitellinum, Cassia corymbosa* and *Asplenium onopteris*, which were frequently observed in the first surveys (upriver), are rare or absent in the last surveys (downriver).

This reduction in species richness downstream would be due to the substitution of certain species by invasive species. There are two possible reasons for this phenomenon. The first is that invasive species increase in proportion to the number of bridges crossing the river, which can act as gateways for new species. The second reason is that the river's waters contribute to the dispersal of the species' propagules. Invasive species thus have a higher ecological plasticity than other species, and establish themselves quickly and easily.

CONCLUSION AND SUGGESTIONS

1. Conclusion

The overall aim of this project, entitled "Characterization of invasive plants in the marshes of the Ruvubu River", was to contribute to knowledge of the invasive plants present in the marshes of the Ruvubu River, with a view to providing Burundian biodiversity managers with a reference tool for the preventive and curative control of invasive plants.

The methodology used to study the flora of our study area is mainly that of the systematic approach based on phytosociological surveys to establish the floristic richness of these marshes. Invasive plants were identified by combining average species cover with invasive plasticity characteristics.

The results show that the marshes of the Ruvubu River are relatively diverse, with 161 species in 125 genera belonging to 54 families. Among these families, only six are the best represented: Poaceae (17.39%), Asteraceae (14.91%), Fabaceae (10.56%), Cyperaceae (4.97%), Malvaceae (3.73%) and Euphorbiaceae (3.11%).

Analysis of biological types revealed a dominance of chamephytes (33.73%), over therophytes (19.79%), geophytes (16.93%), phanerophytes (15.78%) and hemicryptophytes (13.77%).

Analysis of diaspora types (Fig.3.5) shows that anemochores (pogonochores, pterochores and sclerochores) account for 56.37%, autochores (ballochores and barochores) for 23.49% and zoochores (sarcochores and desmochores) for 20%.

Phytogeographically, widespread species (62.04%) are the most abundant, followed by African multi-regional species (19.42%) and regionally distributed species (11.70%), then American species (5.30%) and finally link species (1.54%).

The combination of average species cover and invasive plasticity led to the selection of 13 invasive species: *Centella asiatica*, *Digitaria horizontalis*, *Lantana camara*, *Pennisetum trachyphyllum*, *Galinsonga parviflora*, *Cynodon dactylon*, *Tagetes minuta*, *Eleusine indica*, *Cynodon nemfluensis*, *Ageratum conyzoides*, *Nephrolepis undulata*, *Polygonum glabrum* and *Imperata cylindrica*. The invasive plants selected fall into six botanical families, two of which (Poaceae and Asteraceae) are the most represented, accounting for 69.23% of all invasive species. The floristic gradient of selected invasive plants increases from upstream to downstream of the river

Seven other species *Oxalis latifolia*, *Mimosa diplotricha*, *Mimosa pigra*, *Xanthium strumarium*, *Panicum maximum*, *Acanthospermum australe* and *Paspalum notatum* are considered potential invasive plants in the Ruvubu river marshes.

2. Suggestions

At the end of this work, the floristic composition and invasive plants of the Ruvubu river marshes were highlighted. However, knowledge of these plants is not sufficient for the preventive and curative control of biological invasions in Burundi. For effective management of the latter, further research is essential, in particular :

- studies on invasive species in the Ruvubu National Park and on the current and potential distribution of invasive species already known in Burundi in order to define a management plan;
- studies on invasive species in other marshes and ecosystems (natural, semi-natural or anthropized) in Burundi.

But strategies for effective, concrete action to manage invasive species also need to be drawn up, taking into account the ecological and biological diversity of Burundi's ecosystems.

References

Abram, P. K., & Moffat, C. E. (2018). Rethinking biological control programs as planned invasions. *Current Opinion in Insect Science, 423,* 1-7. https://doi.org/10.1016/j.cois.2018.01.011

Agboola, O., & Joseph, I. M. (2014). Impact of *Tithonia diversifolia* (Hemsly) A. Gray on the soil, species diversity and composition of vegetation in Ile-Ife (Southwestern Nigeria), Nigeria. *International Journal of Biodiversity and Conservation, 6*(7), 555-562. https://doi.org/10.5897/ijbc2013.0634

Akodéwou, A., Johan, O., Sêmihinva, A., Laurent, G., Koffi, A., & Gond, V. (2019). Invasive plant issues in southern Togo (West Africa): contribution of landscape systems analysis and remote sensing. *Biotechnology, Agronomy, Society and Environment*, 23(2), 88-103.

Allen, W. L., Street, S. E., & Capellini, I. (2017). Fast life history traits promote invasion success in amphibians and reptiles. *Ecology Letters,* 20(2), 222-230. doi:10.1111/ele.12728

APRN/BEPB. (2012). Etude de référence environnementale et socio-économique en colline Rabiro, sous-colline Taba, en commune Mutumba, province de Karusi. p.28. unpublished.

Bangirinama, F. (2010). Ecosystem restoration processes during post-cultivation dynamics in Burundi: mechanisms, characterization and ecological series. Thèse de dctorat. Université Libre de Bruxelles.200 p.

Bavumiragiye, F., & Niyonkuru, D. (2018). Contribution à l'étude des espèces végétales envahissantes au Burundi:cas de la vallée de la rivière Murembwe en commune Rumonge. dissertation.ENS.49 p.

Ben-ghabrit, S., Bouhache, M., Birouk, A., & Bon, M.-C. (2018). When invasive alien plants threaten agriculture and ecosystems. Eleventh Congress of the Moroccan Plant Protection Association, May, 33.

Bifolchi, A. (2007). Biology and population genetics of an invasive species: the case of the American mink (*Mustela vison* Schreber , 1777) in Brittany. PhD thesis. Université d'Angers.160 p.

Bizuru, E. (2005). Etude de la flore et de la végétation des marais du Burundi. PhD thesis. Université Libre de Bruxelles.298 p.

Blanfort, V., Fabre, J., Huguein, J., Balent, G., & Daures, S. (2009). Cattle breeding, invasive plants and biodiversity in New Caledonian dry forests. In: *Seizièmes rencontres autour des recherches sur les ruminants*, Paris, December 2-3, 2009. INRA. Paris: Institut de l'élevage, 237-240.

Bousquet, T., Waymel, J., Zambettakis, C., & Geslin, J. (2016). List of invasive vascular plants in Basse-Normandie. DREAL de Normandie / Région de Normandie. Villers-Bocage: Conservatoire botanique national de Brest, 28 p. + appendices

Bousquet, T., Waymel, J., Zambettakis, C., Geslin, J., & Magnanon, S. l v i e. (2013). Liste des plantes vasculaires invasives de Basse-Normandie, Dreal Basse-Normandie/Conseil régional de Basse-Normandie. Villers-Bocage: Conservatoire botanique national de Brest, 39 p.

Bouzillé, J. B. (2007). Gestion des habitats naturels et biodiversité: Concepts, méthodes et démarches. Tec & Doc Lavoisier. 331 p.

Braun-blanquet, J. (1932). Plant sociology: The study of plant communities. Ed. Mac Gray Hill, New York, London, 439 p.

Breton, G. (2014). Introduced or invasive species in the ports of Le Havre d'Antifer and Rouen (Normandy, France). *Hydroécologie Appliquée.* 18, 23-65. https://doi.org/10.1051/hydro/2014003

Byers, J. E., Reichard, S., Randall, J. M., Parker, I. M., Smith, C. S., Lonsdale, W. M., Atkinson, I. A. E., Seastedt, T. R., Williamson, M., Chornesky, E., & Hayes, D. (2002). Directing Research to Reduce the Impacts of Non indigenous Species. *Conservation Biology*, *16*(3), 630-640.

CAB International. (2004). Prevention and Management of Invasive Alien Species: Implementation of Cooperation in West Africa. Proceedings of a Workshop held in Accra, Ghana, March 9-11, 2004 (CAB International Nairobi Kenya (ed.); 115 p.). GISP Secretariat.

Canton of Valais. (2017). Gestion des néophytes envahissantes en Valais : Bilan et plan d ' action 2017-2020. Sion. 44 p.unpublished.

Cassey, P. (2002). Life history and ecology influences establishment success of introduced land birds. *Biological Journal of the Linnean Society*, *76*, 465-480. https://doi.org/10.1046/j.1095-8312.2002.00086.x

Cassey, P., Blackburnw, T. M., Russell, G., Katee, E. J., & Lockwood, J. L. (2004). Influences on the transport and establishment of exotic bird species: an analysis of the parrots (Psittaciformes) of the world. *Global Change Biology*, *10*, 417-426. https://doi.org/10.1111/j.1529-8817.2003.00748.x

Dajoz, R. (2006). Précis d'écologie (DUNOD (ed.); 8th edition). 630 p.

Dansereau, P., & Lems, K. (1957). The grading of dispersal types in plant communities and their ecological significance. Contribution de l'Institut de Botanique de l'Université de Montréal, 71, 1-52.

Direction Régionale de l'Environnement en Guyane. (2010). Biological invasions in French Guiana Biological invasions in French Guiana. 1ère phase: Diagnostic. CIRAD. 144 p.

Dukes, J. S., & Mooney, H. A. (1999). Does global change increase the success of biological invaders? *Trends in Ecology & Evolution*, *14*(4), 135-139.

Duperron, B., Traclet, S., Viscardi, G., Guiot, V., Dimassi, A., Lavergne, C., & Gigord, L.D.B. (2019). Stratégie de lutte contre les espèces végétales invasives à Mayotte: Diagnostic et programme opérationnel de lutte (Issue Version 2.3). Conservatoire Botanique National de Mascarin & DEAL. 87p.

Dushimirimana, S., Masharabu, T., Bizuru, E., & Bigendako, M. J. (2010). Flora and natural vegetation of the Nyamuswaga marshes, Burundi. *INECN Scientific Bulletin*, 8(10-15).

Fischer E, & Killmann D. (2008). Plants of Nyungwe Park-Rwanda (ORTPN). University of Koblenz-Landau.780 p.

Fleriag, L. (2009). Participation in preparatory work to establish the list of potentially invasive exogenous species to be banned from import. dissertation. University of Perpignan. 27 p.

Fourdrigniez, M., & Meyer, J.-Y. (2008). List and characteristics of naturalized and invasive introduced plants in French Polynesia. Contribution à la Biodiversité de Polynésie française N°17. Délégation à la Recherche, Papeete, 62 pages + Appendix.

Fournier, A. (2018). Modeling and predicting biological invasions. PhD thesis. Université Paris-Saclay. 265 p.

Fumanal, B. (2007). Characterization of the biological traits and evolutionary processes of an invasive species in France: *Ambrosia artemisiifolia* L. PhD thesis. Université De Bourgogne. 237 p.

Gérard, B., Didier, A., Vincent, B., & Gibon, A. (1998). Grazing activities, landscapes and biodiversity. *Annales de Zootechnie*, 47,419-429. https://doi.org/10.1051/animres:19980509

Grall, J., & Coic, N. (2006). Synthèse des méthodes d'évaluation de la qualité du benthos en milieu côtier. Rapport de travail. LEMAR. *2005*, 90 p.

Invasive Species Group. (2011). Invasive plants for natural environments in New Caledonia. Agency for the prevention and compensation of agricultural or natural disasters. Nouméa. 224 p.

Guo, Q. (2006). Intercontinental biotic invasions: What can we learn from native populations and habitats? *Biological Invasions*, *8*(7), 1451-1459. https://doi.org/10.1007/s10530-005-5834-1

Habiyaremye, F. M., & Nzigidahera, B. (2016). Habitats du Parc National de la Kibira (Burundi)-Lexique des plantes pour connaître et suivre l'évolution des forêts du secteur Rwegura. Royal Belgian Institute of Natural Sciences (IRSNB). 144p.

Habonayo, R., Azihou, A. F., Dassou, G. H., Hitimana, M., & Cossi, A. (2019). Effect of the invasive liana *Sericostachys scandens* Gilg & Lopr . (Amaranthaceae) on the spatial structure and recruitment of woody plant species in Kibira National Park, Burundi. *International Journal of Innovation and Scientific Research*, *44*(2), 159-170.

Hakizimana, P., Bangirinama, F., Masharabu, T., Habonimana, B., De Cannière, C., & Bogaert, J. (2012). Vegetation characterization of Kigwena dense forest and Rumonge open forest in Burundi. *Bois & Forets Des Tropiques*, 312, 43-52. https://doi.org/10.19182/bft2012.312.a20502

Heger, T., & Trepl, L. (2003). Predicting biological invasions. *Biological Invasions*, *5*(4), 313-321. https://doi.org/10.1023/b:binv.0000005568.44154.12

IUCN-GISD. (2021). http://www.iucngisd.org/gisd/. http://www.iucngisd.org/gisd/

IUCN-PAPACO. (2013). Invasive plants affecting protected areas in West Africa. Management for biodiversity risk reduction. 84 p.

IUCN. (2021). *https://www.iucn.org/content/global-invasive-species-database-gisd).* https://www.iucn.org/content/global-invasive-species-database-gisd

Jacques, T., Christophe, L., Serge, M., Vincent, B., Stéphane, B., Thomas, L. B., Julien, T., & Jean-Noël, R. (2006). Bilan des connaissances sur les conséquences écologiques des invasions de plantes à l'île de la Réunion (archipel des Mascareignes, océan indien). *Revue d'écologie - la Terre et la Vie. 61*, 35-52.

Jean-François, A., Olivier, R., Charlotte, J., Bruno, M., & Virginie, S. (2012). Étude sur les plantes exotiques envahissantes sur des Espaces Naturels Sensibles en Essonne - Cartographie et préconisation de gestion. O.G.E. - Office de Génie Écologique. 104 p.

Joy B., Z., & Kercher, S. (2004). Causes and Consequences of Invasive Plants in Wetlands: Opportunities, Opportunists, and Outcomes. *Critical Reviews in Plant Sciences*, *23:5*, 431- 452. https://doi.org/10.1080/07352680490514673

Julie, L., Martha, H., & Michael, M. (2007). Invasion Ecology. Blackwell Publishing Ltd. Singapore. 313 p. https://doi.org/10.1524/ncrs.2001.216.14.683

LeBorgeois T. Grard, P. & Merlier, H. (1995). Adventrop: une base de connaissance interactive des adventices d'Afrique soudano-sahelienne. *Agriculture et développement*. 8, 51-55

Lebrun, J. P. (1947). The vegetation of the alluvial plain south of Lake Edouard. Institut des Parcs Nationaux du Congo Belge, Exploration du Parc National Albert. Mission Lebrun (1937-1938). Fascicule 1: 472-800. Brussels.

Lee, C. E. (2002). Evolutionary genetics of invasive species. *Trends in Ecology & Evolution*, *17*(No.8), 386-390. https://doi.org/10.1016/S0169-5347(02)02554-5

Lefeuvre, J.-C. (2016). Les invasions biologiques: un risque pour la biodiversité à l'échelle mondiale.Beaulieu.IRD.49 p. https://doi.org/10.4000/books.irdeditions.7656

Lisan, B. (2014). Invasive plants in Madagascar. Version V1.0. Madagascar. 185 p.

Masabo, O., & Nindorera, D. (2019). Biodiversity perception monitoring study based on selected indicators in line with Aichi Goal 1.OBPE. 75 p.

Masharabu, Tatien, Manirakiza, O., Ndayishimiye, J., & Bangirinama, Frédéric Havyarimana, F. (2014). Diversity and conservation of autochthonous woody plants in anthropized landscapes: the case of the Kabuye Zone in Commune Matongo (Burundi). *Bulletin Scientifique de l'Institut National Pour l'environnement et La Conservation de La Nature, 13: 35-42.*

Masharabu, Thatien. (2011). Flora and vegetation of Ruvubu National Park in Burundi: diversity, structure and implications for conservation. PhD thesis. Université Libre de Bruxelles. 224 p.

Masumbuko, C.N. (2011). Ecology of *Sericostachys scandens*, an invasive liana in the mountain forests of Kahuzi-Biega National Park, Democratic Republic of Congo. PhD thesis. Université Libre de Bruxelles. 192 p.

Meddour, R. (2011). La méthodologie phytosociologique sigmatiste ou Braun-blanqueto-tüxenienne. Tizi Ouzou (unpublished). 40 p.

MEEATU. (2014). Atlas of the four Ramsar sites-Location and Resources. Bujumbura. 42 p.

Mendoza, G. H. (2016). Identification of biodiversity loss risks in the face of anthropogenic pressures and climate change to 2100: Application of dynamic conservation to the Alpes-Maritimes territory. Architecture, spatial planning. Ecole Nationale Supérieure des Mines de Paris. 321 p.

Merlier H, & Montegut J. (1982). Tropical weeds: Plant and adult flora of 123 African and pantropical species. MRE-France. 490 p.

MEEATU. (2013). *Stratégie Nationale et Plan d'Action sur la Biodiversité* 2013-2020.

Bujumbura. 104 p.

Mooney, H. A., & Cleland, E. E. (2001). The evolutionary impact of invasive species. *Proceeding of the National Academy of Sciences U.S.A.*, *98*(10), 446-5451.

Muoghalu, J. I., & Chuba, D. K. (2005). Seed germination and reproductive strategies of Tithonia *diversifolia* (Hemsl.) Gray and *Tithonia rotundifolia* (P.M) Blake. *Applied Ecology and Environmental Research*, *3*(1), 39-46. https://doi.org/10.15666/aeer/0301_039046

Mwanyambo, M., Kamwendo, S. J., Patel, H. I., Kathumba, S. E., Wong, J. L., & Pagad, S. (2020). *GRIIS* Checklist of Introduced and Invasive Species - Malawi. Version 1.2. *Invasive Species Specialist Group ISSG. Checklist dataset.* https://doi.org/10.15468/j6du5s accessed via GBIF.org

Ndayisaba, S. (2018). Study of invasive exotic herbaceous plants in the city of Bujummbura already installed in the wild. dissertation. ENS. 50p.

Nduwimana, A. (2014). Characterization of the Malagarazi (Burundi) natural landscape and approach to Sustainable Conservation of its biodiversity. PhD thesis. École doctorale Sciences de la nature et de l'Homme - Évolution et écologie (Paris). 269 p.

Nduwimana, A., Habonayo, R., Ndayizeye, B., & Hitimana, M. (2021). Analyse phytosociologique de la végétation de la réserve naturelle forestière de Vyanda au Sud-Ouest du Burundi Phytosociological analysis of the vegetation of the Vyanda natural forest reserve in southwestern Burundi. *International Journal of Biological and Chemical Sciences* 15(4):1325-1337.

Nzigidahera, B. (2017). Situation of invasive species in Burundi. Bujumbura. 76 p.

Nzigidahera, B., Habiyaremye, F. M., Mbarushimana, D., Masabo, O., Bisthoven, L. J. De, & Habonimana, B. (2020). Habitats du Parc national de la Ruvubu (Burundi)-État actuel et guide au suivi de leur dynamique à l'aide d'un lexique des plantes (H. F. Muhashy (ed.)). Royal Belgian Institute of Natural Sciences (IRSNB). 245 p.

Nzigidahera, B., & Habonimana, B. (2015). Study of the causes of key biodiversity issues in Burundi (CHM-Burundi). OBPE. 44 p.

Osawa, T., Mitsuhashi, H., & Niwa, H. (2013). Many alien invasive plants disperse against the direction of stream flow in riparian areas. *Ecological Complexity*, *15*, 26-32. https://doi.org/10.1016/j.ecocom.2013.01.009

OSS. (2020). Terrestrial invasive alien species (IAS) in the Arab Maghreb: Current status and prospects for a sub-regional strategy (Algeria, Libya, Mauritania, Morocco and Tunisia).

148 p.

Patrick, T. (2017). Encyclopedic dictionary of biological diversity and nature conservation (T. Patrick (ed.); Third). 1056 p.

Piry, S., Alapetite, A., Cornuet, J.-M., Paetkau, D., Baudouin, L., & Estoup, A. (2004). GENECLASS2: A Software for Genetic Assignment and First-Generation Migrant Detection. *Journal of Heredity*, 95(6), 536-539. https://doi.org/10.1093/jhered/esh074

Primack, R. B., Sarrazin, F., & Lecompte, J. (2012). Conservation Biology. Dunod. 351p.

Quéré, E., Ragot, R., Geslin, J., Magnanon, S., & Haury, J. (2011). List of invasive vascular plants in Brittany. *CSRPN de Bretagne*, 24 p.

Ramade, F. (2009). Éléments d'Écologie-Écologie appliquée : action de l'Homme sur la biosphère.7ème Edition. Dunod. 789 p.

Raunkiaer, C. (1934). The life's forms of plants and statistical plant geography. London. 632 pp.

Reekmans, M., & Niyongere, L. (1983). Vernacular lexicon of vascular plants of Burundi. U.B. 58 p.

Rejmánek, M., Richardson, D. M., & Pyšek, P. (2013). Plant Invasions and Invasibility of Plant Communities. *Vegetation Ecology* (2),387-424.

Richardson, D. M., & Pyšek, P. (2007). The ecology of invasions by animals and plants. *Progress in Physical Geography*, *31*(6), 659-666. https://doi.org/10.1177/0309133307087089

Ricklefs, R. E., & Miller, G. L. (2005). Ecologie (4th Edition). De Boeck Supérieur. 858 p.

Soubeyran, Y. (2010). Managing invasive alien species. IUCN French Committee. 68 p.

Soubeyran, Yohann. (2008). Espèces exotiques envahissantes dans les collectivités françaises d ' outre-mer Etat des lieux et recommandations (Collection Planète Nature). IUCN French Committee, Paris, France. 202 p.

Tassin, J. (2002). Dynamiques et conséquences de l'invasion des paysages agricoles des Hauts de la Réunion par *Acacia mearnsii* De Wild. PhD thesis. Université Paul Sabatier. 215 p.

Teyssèdre, A., & Couvet, D. (2010). Écologie et biodiversité: Des populations aux socioécosystemes (Belin). 336 p.

Toussaint, B., Lambinon, J., Dupont, F., Verloove, F., Petit, D., Hendoux, F., Mercier, D., Housset, P., Truant, F., & Decocq, G. (2007). Reflections and definitions relating to the native or introduced status of plants; application to the flora of northwestern France. *Acta Botanica Gallica*, 154 (4), 511-522. https://doi.org/10.1080/12538078.2007.10516077

Troupin, G. (1978). Flora of Rwanda: spermatophytes (vol. 1). Musée Royal de l'Afrique Centrale. 413 p.

Troupin, G. (1983). Flora of Rwanda: spermatophytes (Vol. 2). Musée Royal de l'Afrique Centrale. 603 p.

Troupin, G. (1985). Flore du Rwanda: spermatophytes (Vol.3). INRS. 729 p.

Troupin, G. (1988). Flora of Rwanda: spermatophytes (Vol.4). Musée Royal Afrique Centrale. 651 p.

IUCN/PACO. (2013). Invasive plants affecting protected areas in West Africa: management for biodiversity risk reduction. IUCN/PACO, Ouagadougou, Burkina Faso. 84 p.

Valéry, L., Hervé, F., Jean-claude, L., & Daniel, S. (2008). In search of a real definition of the biological invasion phenomenon itself. *Bioloical Invasions*, 10(8), 1345-1351. https://doi.org/10.1007/s10530-007-9209-7

Van Der Plas, F., Manning, P., Soliveres, S., Allan, E., Scherer-Lorenzen, M., Verheyen, K., Wirth, C., Zavala, M. A., Ampoorter, E., Baeten, L., Barbaro, L., Bauhus, J., Benavides, R., Benneter, A., Bonal, D., Bouriaud, O., Bruelheide, H., Bussotti, F., Carnol, M., ... Schlesinger, W. H. (2016). Biotic homogenization can decrease landscape-scale forest multifunctionality. *Proceedings of the National Academy of Sciences of the United States of America*, *113*(13), 3557-3562. https://doi.org/10.1073/pnas.1517903113

Van Kleunen, M., Weber, E., & Fischer, M. (2010). A meta-analysis of trait differences between invasive and non-invasive plant species. *Ecology Letters*, *13*(2), 235-245. https://doi.org/10.1111/j.1461-0248.2009.01418.x

Vanderhoeven, S., Branquart, E., Grégoire, J.-C., Mahy, G., Brahy, O., Gilbert, M., La Spina, S., Nicolas, M., Frédéric, P., & Pieret, N. (2006). Invasive alien species: Dossier scientifique réalisée dans le cadre de l'élaboration du Rapport analytique 2006-2007 sur l'état de l'environnement wallon. 42 p.

Vanessa, H., Mickaël, le C., Frédéric, R., & Vincent, B. (2009). Invasive alien species in New Caledonia. IRD. 87 p.

Vitousek, P. M., M D'antonio, C., Loope, L. L., Rejmánek, M., & Westbrooks, R. (1997). Introduced species: A significant component of human-caused global change. *New Zealand Journal of Ecology*, *21 (1)*, 1-16.

White, F. (1979). The Guineo-Congolian Region and Its Relationships to Other Phytochoria. *Bulletin Du Jardin Botanique National de Belgique*, 49, No. 1/, 11-55. https://doi.org/10.2307/3667815

White, F. (1983). Vegetation of Africa: A Descriptive Memoir to Accompany the

Unesco/AETFAT/UNSO Vegetation Map of Africa. UNESCO/AETFAT/UNSO, ORSTOM-UNESCO. 356 p.

Williamson, M. (2006). Explaining and predicting the success of invading species at different stages of invasion. *Biological Invasions*, *8*(7), 1561-1568. https://doi.org/10.1007/s10530-005-5849-7

Witt, A., Wong, L. J., & Pagad, S. (2020). Global Register of Introduced and Invasive Species - Tanzania. Version 1.2. Invasive Species Specialist Group ISSG. Checklist dataset. GBIF.org. https://doi.org/10.15468/7n4vid accessed via GBIF.org

Woods, K. D. (1997). Community Response to Plant Invasion. Assessment and Management of Plant Invasions. Springer Series on Environmental Management. 56–68. https://doi.org/10.1007/978-1-4612-1926-2_6

Zihalirwa, B., Mutabana, N., & Lejoly, J. (2020). Ecological study of the invasive liana Sericostachys scandens in the high-altitude part of the Kahuzi -Biega National Park (PNKB), (Sud - Kivu, R.D.Congo). *Annales Des Sciences. Université Officielle de Bukavu*, 1(1), 28-36.

Appendix s

Appendix 1. Phytosociological table of surveys

				Sites/Collines échantillonnés	Migende	Mpangu	Kinga	Musonge	Shinya	Shurugumya	Gahise	Nyarurama	Nkango	Gatura 1	Jimbi	Gatura 2	Gisara	Kibimba	Rushubi	Buniha	Gitamo	Nyangungu 1	Ruyaga	Nyangungu 2	Munanira 1	Rurengera	Munanira 2	Ruyenzi	Muharuro	Mitobo	Shinyanga	Kinzerere	Bihembe	Rutonganikwa	Kiguruka	Kaziga	Bukwavu	Nyagisenyi	Nyarusange	Wangugo	Kiryama	Gisagara
	TB	TP	TD	Espèces	R1	R2	R3	R4	R5	R6	R7	R8	R9	R10	R11	R12	R13	R14	R15	R16	R17	R18	R19	R20	R21	R22	R23	R24	R25	R26	R27	R28	R29	R30	R31	R32	R33	R34	R35	R36	R37	R38
1	P	Afr-Trop	Ballo	*Acacia polyacantha* Willd.	0	0	0	0	0	0	0	0	0	0	0	0	0	0	0	0	1	0	0	0	0	0	0	0	0	0	0	0	0	1	1	0	1	0	1	0	1	1
2	P	SZ(O)	Ballo	*Acacia sieberana* DC.	0	0	0	0	0	0	0	0	0	0	0	0	0	0	0	0	0	0	0	0	0	0	0	0	0	0	0	0	0	0	0	0	1	1	0	1	1	1
3	Th	Am	Desmo	*Acanthospermum australe* (Loefl.) Kuntze	0	0	0	0	0	0	0	0	0	0	0	0	3	3	2	2	3	4	0	3	2	0	3	0	0	0	0	0	0	0	0	0	0	0	0	0	0	0
4	P	Afr-Trop	Ballo	*Acanthus polystachyus* Del.	0	2	2	0	0	3	3	2	2	0	2	0	2	2	0	0	1	1	0	1	0	1	1	0	3	3	0	3	0	2	2	0	2	0	0	0	2	3
5	Ch	Pan	Desmo	*Achyranthes aspera* L	2	2	0	0	0	2	2	0	1	1	0	2	2	2	0	0	2	2	2	0	0	0	0	0	0	0	0	0	0	0	0	0	0	0	0	2	3	3
6	Th	Subcos	Pogo	*Ageratum conyzoides* L	2	2	0	2	2	5	3	3	2	2	2	3	2	3	3	3	3	3	3	3	3	3	3	3	3	3	3	3	3	2	3	3	2	2	2	2	2	2
7	P	Afr-Mal	Ballo	*Albizia gummifera* (J.F.Gmel.) C.A.Sm.	0	0	0	0	0	0	0	0	0	0	0	0	0	0	0	0	0	0	0	0	0	0	0	0	0	0	0	0	0	1	0	0	0	0	0	0	0	0
8	Th	SG	Scléro	*Amaranthus viridis* L	2	2	0	0	0	0	0	0	0	0	0	0	0	0	0	0	2	2	0	0	0	0	0	2	0	0	2	0	0	0	0	0	0	0	1	0	0	0
9	H	SZ(O)	Scléro	*Aristida adoensis* Hochst. ex A.Rich.	0	0	0	0	0	0	0	0	0	0	0	0	0	0	0	0	2	2	0	0	0	0	0	0	0	0	0	0	0	2	2	0	0	0	0	0	0	0
10	Ge	Mont	Scléro	*Arthropteris anniana* Lawalrée	0	0	0	0	0	0	0	0	0	0	0	0	0	0	0	0	2	0	0	0	0	0	0	0	0	0	0	0	0	2	0	0	0	0	0	0	3	0
11	Ge	Afr-Trop	Scléro	*Arthropteris orientalis* (J.F. Gmel.)	0	0	2	2	2	2	0	2	2	0	0	2	2	0	0	2	3	2	2	2	2	2	2	0	0	2	0	2	2	2	2	2	2	0	0	0	0	3
12	Ge		Scléro	*Arundinalia alpina* K.Schum.	2	0	0	0	0	0	0	0	0	0	0	0	0	0	0	0	0	0	0	0	0	0	0	0	0	0	0	0	0	3	0	0	0	0	0	0	3	0
13	Th	Plur Afr	Pogo	*Aspilia pluriseta* Schweinf.	2	2	0	1	0	2	2	0	2	0	2	2	2	0	1	0	0	0	0	0	0	0	0	0	0	0	0	0	0	0	0	0	0	0	0	0	0	0
14	Ge		Scléro	*Asplenium onopteris* L.	2	2	2	2	2	2	2	2	2	2	2	2	2	2	2	2	2	2	0	0	0	0	0	0	0	0	0	0	0	0	0	0	0	0	0	0	0	0
15	Th	Pan	Desmo	*Bidens pilosa* L	2	2	2	2	2	2	2	2	2	2	2	2	2	2	2	2	2	2	2	2	2	2	2	2	2	2	2	2	2	2	2	2	2	2	2	2	2	2
16	Th	LSZ-Mo	Ballo	*Biophytum helenae* Buscal. & Muschl.	0	1	0	0	1	0	1	0	0	0	0	0	0	0	0	1	0	0	0	0	0	1	0	0	0	0	0	0	1	0	0	0	0	1	0	0	0	0
17	Th	Cos	Ptéro	*Blumea brevipes* (Oliv. & Hiern) Wild	0	0	0	0	0	0	1	0	0	0	0	0	0	0	0	1	0	0	0	0	0	0	0	0	0	0	0	0	0	0	0	0	0	0	0	0	0	0
18				*Bulbostylis densa* (Wall.) Hand.-Mazz.	0	0	0	0	0	0	0	0	0	0	0	0	0	0	0	0	0	0	0	0	0	0	0	0	0	0	0	0	0	2	2	0	0	0	0	0	0	0
19	Hy	Afr-Trop		*Burnatia enneandra* (Hochst.) Micheli	0	0	0	0	0	3	0	0	0	0	0	0	0	0	0	0	0	0	0	0	0	0	0	0	0	0	0	0	0	0	0	0	0	0	0	0	0	0
20				*Capsella bursa-pastoris* (L.) Medik.	0	0	0	0	0	0	0	0	0	0	0	0	0	0	0	0	2	2	0	0	0	0	0	0	0	0	0	0	0	0	0	0	0	0	0	0	0	0
21	P	Pal		*Carissa spinarum* L	0	0	0	0	0	0	0	0	0	0	0	0	0	0	0	0	3	3	2	2	2	0	2	2	0	0	2	2	2	3	2	2	2	2	2	2	2	3
22				*Cassia corymbosa* Lam.	2	2	2	0	0	0	0	2	2	2	0	0	0	0	0	0	0	0	0	0	0	0	0	0	0	0	0	0	0	0	0	0	0	0	0	0	0	0
23	P	SZ	Ballo	*Cassia didymobotrya* Fres.	0	0	0	0	0	0	0	0	0	0	0	2	2	2	2	0	2	2	0	0	2	0	0	0	2	0	0	0	0	2	2	0	0	0	0	2	3	2
24	T	Pan		*Cassia occidentalis* L.	0	0	0	0	0	0	0	0	0	0	0	0	2	2	0	0	2	2	2	0	2	0	2	0	2	2	0	2	0	2	2	0	2	0	0	2	2	2
25				*Cenchrus unisetus* (Nees) Morrone	0	0	0	0	0	0	0	0	0	0	0	0	2	0	2	2	3	3	3	2	0	2	2	0	2	2	2	2	2	3	3	3	2	2	2	2	2	2
26	Ch	Pal	Ballo	*Centella asiatica* (L.) Urb.	4	3	3	2	2	2	3	3	4	3	3	3	4	3	3	4	4	3	4	4	3	3	3	3	4	3	3	3	4	3	3	4	3	3	3	3	3	4
27				*Chenopodium ugandae* Aellen	0	0	0	0	0	0	0	0	0	0	0	0	0	0	0	0	0	0	0	0	0	0	0	0	0	0	0	2	2	0	0	0	2	0	0	2	2	0
28	Ch	Afr-Trop	Sarco	*Cissampelos mucronata* A.Rich.	0	2	2	2	2	2	2	2	2	2	2	2	2	2	2	2	2	2	2	2	2	2	2	2	2	2	2	2	2	2	2	2	2	2	2	2	2	2
29	P	Mont (Afr-Trop)		*Clerodendrum johnstonii* Oliv.	0	0	0	2	2	2	2	0	0	0	0	2	2	2	2	0	2	2	0	2	2	2	2	0	0	0	2	2	2	2	2	2	0	0	2	2	2	2
30	P	SZ	Ptéro	*Combretum collinum* Fresen.	0	0	0	0	0	0	0	0	0	0	0	0	0	0	0	0	2	2	2	2	2	2	2	2	2	2	2	2	2	2	2	2	2	2	2	2	2	2
31	Ch	Plur-Afr	Ballo	*Commelina africana* L	0	2	2	2	2	2	2	2	2	2	3	3	3	3	3	0	0	3	3	0	1	0	1	0	3	3	3	0	3	3	0	3	3	0	0	3	2	0
32	Ch	Pal	Ballo	*Commelina benghalensis* L	3	0	3	0	0	3	0	3	0	0	0	2	2	2	2	0	2	2	2	0	2	3	3	2	0	0	3	3	0	2	0	2	2	2	2	2	3	3
33	Th	Pal	Pogo	*Conyza aegyptiaca* (L.) Ait.	0	0	0	0	0	2	2	2	2	2	0	0	2	0	0	0	2	2	0	0	0	0	0	0	0	0	2	2	2	2	2	2	2	2	2	0	2	0
34	P	Plur-Afr	Sarco	*Cordia africana* Lam.	0	0	0	0	0	0	0	0	0	0	0	0	0	0	0	0	0	0	0	0	0	0	0	0	0	0	0	0	0	0	0	0	0	0	0	0	2	0
35	Th	Afr-Mal	Pogo	*Crassocephalum crepidioides* (Benth.) S. Moore	1	1	0	0	0	2	0	2	2	2	0	0	0	0	0	0	2	2	0	0	0	2	2	0	0	0	0	0	0	2	2	2	0	0	0	2	2	2
36	Th	Afr-Mal	Pogo	*Crassocephalum montuosum* bumbense S. Moore	0	0	0	0	0	0	0	0	0	0	0	0	0	0	2	2	2	2	2	0	0	0	0	0	2	0	0	0	0	2	2	2	2	2	2	2	2	2
37	Th	Afr-Trop	Pogo	*Crassocephalum Multicorymbosum* S.Moore	0	0	0	0	0	0	0	0	0	0	0	0	2	0	0	0	1	0	0	0	0	0	1	0	0	0	1	0	0	0	0	0	0	0	0	0	0	0
38	Th	Mont	Pogo	*Crassocephalum vitellinum* (Benth.) S. Moore.	2	0	2	2	0	2	2	3	3	3	0	0	0	0	0	0	0	0	0	0	0	0	0	0	0	0	0	0	0	0	0	0	0	0	0	0	0	0
39	Ch	Pal	Ballo	*Crotalaria pallida* Aiton	2	2	2	2	2	2	2	2	2	2	2	2	0	0	0	2	2	2	2	2	0	0	0	0	0	0	0	0	2	2	3	3	0	3	2	2	3	2
40	Th(Ch)	LSZ-G	Ballo	*Crotalaria pumila* Ort. Chipil.	2	2	2	2	2	2	2	2	0	0	2	2	2	2	0	0	2	2	2	0	0	0	2	2	2	2	2	0	0	2	0	3	2	3	2	2	3	3

41	Ch	Pan	Ballo	*Crotalaria retusa* L.	2	2	2	0	0	0	0	0	0	0	0	0	2	2	0	0	0	0	0	0	0	0	0	0	0	0	0	0	0	0	0	0	0	0	0	0	0	0
42	Hc	Pal	Scléro	*Cymbopogon giganteus* Chiov.	0	0	0	0	0	0	0	0	0	0	0	0	0	0	0	0	2	2	2	2	0	0	0	0	0	0	0	0	2	3	3	0	0	2	3	3	2	3
43				*Cynanchum schistoglossum* Schlt	0	0	0	0	0	0	0	0	0	0	0	1	1	1	0	0	0	0	0	0	0	0	0	0	0	0	0	0	0	0	0	0	0	0	0	0	0	0
44	Ch	Subcos	Scléro	*Cynodon dactylon* (L.) Pers.	2	4	2	2	2	0	2	2	2	3	3	3	4	4	2	2	2	2	2	2	3	3	3	3	3	3	3	3	3	2	2	3	3	3	3	4	4	4
45	Ch	Subcos	Scléro	*Cynodon nemfluensis* Vanderyst .	3	2	3	3	3	2	2	2	3	3	3	3	4	3	3	3	2	2	3	3	2	2	3	3	3	3	3	3	3	2	2	2	3	3	3	3	3	3
46	Ge	Plur-Afr	Scléro	*Cyperus articulatus* L.	2	2	2	2	2	2	2	2	2	2	2	2	2	2	2	2	2	2	2	2	2	2	2	2	2	2	2	2	2	2	2	2	2	2	2	2	2	2
47	Ge	Plur-Afr	Scléro	*Cyperus cyperoides* (L.) Kuntze	2	2	2	2	2	2	2	2	2	2	2	2	2	2	2	2	2	2	2	2	2	2	2	2	2	2	2	2	2	2	2	2	2	2	2	2	2	2
48	Hc (Ge)	Pan	Scléro	*Cyperus distans* Linnaeus f.	2	2	2	2	2	3	3	2	2	2	2	3	3	3	2	2	3	3	2	2	2	2	2	2	2	2	2	2	3	3	3	2	2	2	2	2	3	2
49	Ge	Afr-Mal	Scléro	*Cyperus latifolius* Poir.	0	0	0	0	0	2	2	0	0	0	0	2	2	2	2	0	2	2	0	0	0	0	0	0	0	0	0	0	0	2	2	0	0	0	0	0	2	2
50	Ge	Cos	Scléro	*Cyperus papyrus* L.	1	1	0	0	0	0	0	0	2	2	0	0	1	0	0	0	1	0	0	0	0	0	0	0	0	0	0	0	0	1	0	0	0	0	0	0	1	0
51	Ge	Afro-Trop	Sarco	*Cyphostemma adenocaule* (Steud. ex A. Rich.)	0	0	0	0	0	2	2	0	0	0	0	2	2	0	1	1	2	1	0	0	0	0	0	0	0	0	0	2	2	2	0	0	0	0	0	0	0	0
52	Th	Cos	Sarco	*Datura stramonium* L.	1	0	0	0	0	1	1	0	0	0	0	0	0	0	0	0	0	0	0	0	0	0	0	0	0	0	0	0	0	0	0	0	0	0	0	0	0	0
53				*Dichrocephala bicolor* (Roth) Schltdl.	3	3	2	2	2	1	2	0	0	0	2	2	2	0	2	0	3	3	2	0	0	0	0	0	0	0	0	0	2	2	0	0	0	2	0	0	2	2
54	Ge	Pal	Scléro	*Digitaria abyssinica* (Hochst. ex A.Rich.) Stapf	4	3	3	2	2	2	3	2	2	2	2	3	3	4	3	0	0	0	0	0	0	1	0	0	0	0	0	0	0	0	0	0	0	0	0	1	2	4
55	Ge	Pan	Scléro	*Digitaria horizontalis* Willd.	2	3	3	4	4	4	2	3	3	3	0	0	3	2	2	3	4	4	4	4	0	0	3	3	3	3	3	3	3	4	4	4	4	3	0	3	2	2
56	Ge	Pal	Baro	*Dioscorea alata* L.	1	0	0	0	0	1	0	0	0	0	0	0	0	0	0	0	0	2	0	0	0	0	0	0	0	0	0	0	0	0	0	0	0	0	0	0	2	0
57	P	SZ(O)-Mo	Sarco	*Dissotis trothae* Gilg.	2	2	2	2	2	2	2	2	2	2	2	2	0	0	1	1	2	2	2	0	0	2	2	0	0	0	0	0	0	2	2	2	2	2	2	0	0	2
58	P	SZ	Ballo	*Dombeya buettneri* K.Schum.	2	1	1	2	0	2	2	2	2	2	2	2	0	0	0	0	0	0	0	0	0	0	0	0	0	0	0	0	0	0	0	0	0	0	0	0	0	0
59	P	SZ(EOZ)	Sarco	*Dracaena steudneri* Engl.	1	0	0	1	0	0	0	0	1	0	0	0	0	0	0	0	0	0	0	0	0	0	0	0	0	0	0	0	0	0	0	0	0	0	0	0	0	0
60	Th	Pan		*Drymaria cordata* (L.) Willd. ex Roem. & Schult.	3	2	2	2	2	1	1	1	1	1	1	1	1	1	0	0	2	0	0	0	0	0	0	0	0	0	0	0	0	0	0	0	0	0	0	0	0	0
61	Ch	Pal	Scléro	*Eleusine indica* (L.) Gaertn.	2	2	2	2	2	3	3	3	3	3	3	3	3	3	3	3	0	3	3	3	3	3	3	3	3	0	3	3	3	3	3	3	3	3	3	3	3	3
62	Th	L-SZ-Mo	Pogo	*Emilia caespitosa Oliv.*	2	2	2	2	2	3	3	3	3	2	2	1	1	0	0	0	0	0	0	0	0	0	0	0	0	0	0	0	0	0	0	0	0	0	0	0	0	0
63	Ch	SZ(OZ)	Scléro	*Eragrostis olivacea* K.Schum.	0	0	0	0	0	0	0	0	0	0	0	0	0	0	0	0	2	2	0	0	0	1	0	1	0	0	0	0	0	2	2	0	2	0	2	0	0	0
64	Hc	Afr-Mal	Scléro	*Eragrostis tenuifolia* (A.Rich.) Hochst. ex Steud.	1	0	0	1	0	1	0	0	0	0	0	0	2	2	2	2	2	2	2	2	2	2	2	2	2	2	2	2	2	3	3	3	3	3	3	3	3	3
65	Ch	Mont(EA-A	Ballo	*Eriosema montanum* Baker f.	0	0	3	2	2	3	2	0	1	1	1	1	2	2	2	2	0	0	0	0	0	0	0	0	0	0	0	0	0	0	0	0	0	0	0	0	0	0
66				*Erlangea spissa* S. Moore	0	0	0	0	0	2	2	2	2	2	2	2	3	3	2	2	2	2	2	2	2	2	2	2	2	2	3	3	3	2	2	2	2	2	3	3	3	3
67	Th	Am	Scléro	*Euphorbia heterophylla* L.	0	0	0	0	2	2	2	2	2	2	2	2	0	0	2	2	2	2	2	2	0	0	2	0	2	0	0	2	0	0	0	0	2	2	2	2	2	2
68	Ch	Pan	Scléro	*Euphorbia hirta* L.	0	0	0	0	0	0	0	0	0	0	0	0	0	0	0	0	1	1	2	2	2	2	2	2	2	2	0	2	2	2	0	2	2	2	2	2	2	2
69	P	SZ(OZ)	Ballo	*Euphorbia tirucalli* L.	2	0	2	2	2	0	2	0	0	2	2	2	2	2	2	0	0	0	0	0	0	0	0	0	0	0	0	0	0	0	0	0	0	0	0	0	0	0
70	P	LSZ-G	Sarco	*Ficus vallis-choudae* L.	0	0	0	0	2	2	2	2	2	2	0	2	2	0	2	2	0	0	0	0	0	0	0	0	0	0	0	0	0	0	0	0	0	0	0	0	0	0
71	Th	Cos	Desmo	*Galinsoga quadriradiata* Ruiz & Pav.	0	0	0	0	0	0	0	0	0	0	0	0	0	0	0	0	0	0	0	1	1	1	1	1	1	1	1	3	3	3	3	3	3	2	2	2	2	2
72	Th	Cos	Desmo	*Galinsonga parviflora* Cav.	3	3	3	3	3	4	4	4	4	4	3	3	2	2	2	2	2	3	3	3	3	3	3	3	3	3	3	2	2	2	2	2	2	3	3	3	3	3
73	Hc	Plur-Afr	Scléro	*Guizotia scabra* (Vis.) Chiov.	1	2	2	2	2	3	3	3	3	2	1	1	1	1	1	1	0	0	0	0	0	0	0	0	0	0	0	0	0	0	0	0	0	0	0	0	0	0
74	Th	Pal		*Gynandropsis gynandra* (L.) Briq.	1	1	1	1	2	2	2	0	2	0	2	2	2	2	2	2	2	2	2	2	2	0	0	0	0	0	0	0	0	0	0	0	0	0	0	0	0	0
75	P	Afr-Mal	Sarco	*Harungana madagascariensis* Lam. & Poir	0	0	2	2	0	0	0	0	0	0	0	0	0	0	0	0	0	0	0	0	0	0	0	0	0	0	0	0	0	0	0	0	0	0	0	0	0	0
76	Hc	SZ(EOZ)	Pogo	*Helichrysum keilii* Moeser	0	0	0	0	0	0	0	0	0	0	1	0	0	0	0	0	0	0	0	0	0	0	1	1	0	0	0	0	0	0	0	0	0	0	0	0	0	0
77	Th	Plur-Afr	Pogo	*Helichrysum odoratissimum* (L.) Sweet.	2	0	0	0	0	0	2	2	0	0	0	0	0	2	0	2	3	3	2	2	2	3	3	3	2	2	2	2	2	2	2	2	2	2	2	2	2	2
78	Ch	SZ(EOZ)	Desmo	*Hibiscus fuscus* Garcke	0	0	0	0	0	0	0	0	0	0	0	0	0	0	0	0	0	0	0	0	0	0	0	0	0	0	0	0	2	2	0	0	2	0	1	0	0	0
79	Ch	SZ(EOZ)	Desmo	*Hibiscus noldeae* Baker f.	3	1	0	1	0	1	0	0	0	0	0	0	0	0	0	0	0	0	0	0	0	0	0	0	0	0	0	0	0	0	0	0	0	0	0	0	0	0
80				*Hygrophila auriculata* (Schumach.) Heine	1	1	1	2	2	2	2	2	0	0	0	0	0	0	0	0	0	0	0	0	0	0	0	0	0	0	0	0	0	0	0	0	0	0	0	1	1	1
81	P	SZ	Sarco	*Hymenodictyon floribundum* (Hochst. & Steud.) B.L.Rob	0	0	0	0	0	0	0	0	0	0	0	0	0	0	0	0	0	0	0	0	0	0	0	0	0	0	0	2	2	2	2	2	2	2	0	0	0	0
82	Hc	Subcos	Scléro	*Hyparrhenia dichroa* (Steud.) Stapf.	0	0	0	0	0	0	0	0	0	0	0	0	0	0	0	3	3	3	3	2	2	2	2	2	2	2	2	2	3	3	3	3	3	2	2	3	3	4
83	Ch	Plur Afr	Ballo	*Hypoestes triflora* (Forssk.) Roem. & Schult.	2	2	2	2	2	2	2	2	2	2	2	2	3	3	2	2	3	2	2	2	2	3	3	3	3	3	3	3	2	2	2	2	2	2	0	0	0	0
84	Ge	Subcos	Scléro	*Imperata cylindrica* (L.) Beauv.	0	0	0	0	3	3	3	0	0	3	3	3	3	3	3	3	3	3	0	0	3	3	3	3	3	0	0	3	3	3	3	3	0	2	0	2	2	3
85	Ch	Pal	Ballo	*Indigofera spicata* Forssk.	0	0	0	0	0	0	0	0	0	0	0	0	2	2	0	2	2	2	0	0	0	0	0	0	0	0	0	0	0	0	0	0	0	0	2	2	2	2
86	Ch	Pal	Ballo	*Ipomea cairica* (L.) Sweet	0	0	0	0	0	0	0	0	0	0	0	0	3	3	0	0	2	2	2	2	2	2	2	2	2	2	2	2	0	0	0	0	0	0	3	3	3	3
87	Ch	Subcos	Ballo	*Ipomea involucrata*	2	2	2	2	2	2	2	2	2	2	3	3	3	3	0	3	3	2	0	0	2	0	2	2	2	2	3	3	3	0	3	0	2	2	2	2	2	3
88	Ch	Plur Afr	Ballo	*Justicia heterocarpa* T.Anderson.	0	0	0	0	0	0	0	0	0	0	0	0	3	0	0	0	0	0	2	2	2	0	2	2	0	2	2	2	2	2	2	0	0	0	2	2	2	2
89				*Kalanchoe beniensis* De Wild.	0	0	0	0	0	0	0	0	0	0	0	0	0	0	0	0	0	0	0	0	0	0	0	0	0	0	0	0	0	0	0	0	0	0	0	2	2	2
90	P	SZ(OZ)	Ballo	*Kotschya strigosa* (Benth.) Dewit & P.A. Duvign.	0	0	0	0	0	0	0	0	0	0	0	0	0	0	0	0	0	0	0	0	0	0	0	0	0	0	0	0	0	0	0	0	0	1	1	1	2	2
91	Ge	SZ(OZ)	Scléro	*Kyllinga erecta* Schumach.	3	0	2	2	0	2	2	0	0	0	2	2	2	2	2	2	3	2	0	2	2	2	0	2	2	2	2	0	0	2	2	0	0	2	2	0	2	2
92	Hc			*Lactuca serriola* L.	1	1	1	0	0	2	0	0	0	1	0	1	2	0	0	0	0	0	0	0	0	0	1	0	0	1	0	0	0	2	0	0	0	0	2	0	2	2
93	P	Intr	Sarco	*Lantana camara* L.	4	3	3	0	3	3	0	0	0	3	3	0	3	3	0	0	3	3	0	0	0	0	3	3	3	0	4	4	4	4	4	4	0	4	4	4	3	4
94	Th	Pan	Scléro	*Leonotis nepetifolia* (L.) R. Br.	2	0	0	0	0	1	0	0	0	0	0	0	1	0	0	0	0	0	0	0	0	0	0	0	0	0	0	0	0	1	0	0	0	0	0	0	2	1
95	Th	Pan	Scléro	*Leucas martinicencis* (Jacq.) R. Br.	0	0	2	0	0	0	0	0	0	0	0	0	0	0	2	0	2	0	0	0	0	2	2	0	0	0	2	0	2	0	0	0	0	2	2	0	0	2
96	Ch	Plur-Afr	Pogo	*Ludwigia abyssinica* A. Rich.	3	3	0	2	2	2	2	2	2	2	2	2	2	2	2	2	3	2	2	3	2	2	2	3	3	3	2	2	3	3	3	2	3	3	2	2	2	2
97	P	Plur-Afr	Sarco	*Maesa lanceolata* Forsskal	0	2	0	0	0	0	0	0	2	2	2	0	2	2	0	0	3	3	0	3	3	0	0	0	0	0	0	0	3	3	3	0	0	2	2	2	0	0
98	P	Afr-Trop	Sarco	*Maesopsis eminii* Engl.	1	0	0	0	0	0	0	0	0	0	0	1	1	0	0	2	2	2	0	0	0	2	0	0	0	2	0	2	0	0	0	0	0	0	0	0	0	0
99	Ge	Pal	Scléro	*Mariscus sumatrensis* (Retz.) J. Raynal.	2	2	2	3	3	3	3	3	2	2	2	2	2	3	2	2	2	2	2	2	2	2	2	2	2	2	2	2	2	2	2	2	2	2	2	2	2	2
100	P	Afr-Mal	Ballo	*Maytenus arbutifolia* (Hochst. ex A.Rich.) R.Wilczek	0	0	0	0	0	0	0	0	0	0	0	0	0	0	0	0	3	3	2	2	0	2	2	2	2	2	2	3	3	3	3	2	2	2	2	3	3	3

101	Hc	Pan	Scléro	*Melinis minutiflora* P.Beauv.	0	0	0	0	0	0	0	0	2	0	0	0	3	3	0	0	2	2	0	0	0	0	0	2	2	0	0	0	0	0	0	0	2	0	2	0	0	0
102	Hc	pan	Scléro	*Melinis repens* (Willd.) Zizka	0	2	0	2	2	3	3	2	2	2	2	2	2	2	2	0	0	2	2	0	0	2	2	2	0	2	0	2	0	0	0	2	2	2	2	2	2	3
103	P	Pal	Pogo	*Microglossa pyrifolia* (Lam.) Kuntze	0	0	2	2	2	2	0	2	0	2	0	2	2	2	2	0	0	0	0	0	0	0	0	0	0	0	0	0	0	0	0	0	0	0	0	0	0	0
104	Hc	Pan	Ballo	*Mimosa diplotricha* C.Wright.	0	0	0	0	0	0	0	0	0	0	0	0	4	3	0	0	0	0	0	0	3	0	3	0	3	0	0	0	0	4	2	0	0	3	0	0	5	2
105	P(Ch)	Pan	Ballo	*Mimosa pigra L*	0	0	0	0	0	0	0	0	0	0	0	0	0	0	0	0	0	0	0	0	0	0	0	0	0	0	0	0	0	0	0	0	0	0	2	2	2	5
106	Th	Afr-Trop	Ballo	*Momordica foetida* Schumach. & Thonn.	1	1	0	0	0	0	0	0	0	0	0	0	0	0	0	0	0	0	0	2	2	0	0	0	0	0	0	0	0	0	0	0	0	0	0	0	2	0
107				*Morus indica* L.	2	0	0	0	0	0	0	0	0	0	0	0	0	0	0	0	0	0	0	0	0	0	0	0	0	0	0	0	0	0	0	0	0	0	0	0	0	0
108	Ge	Afro-Trop	Ballo	*Neorautanenia mitis* (A.Rich.) Verdc.	0	0	0	0	0	0	0	0	0	0	0	0	3	0	0	0	3	0	2	0	2	0	0	2	0	0	0	2	0	2	0	2	2	0	0	0	0	0
109	Ge		Scléro	*Nephrolepis undulata* (Afzel. ex Sw.) J. Sm.	3	3	3	3	3	3	3	3	3	3	3	3	3	3	3	3	3	3	3	2	2	2	2	3	2	2	2	3	3	2	2	3	3	2	2	2	2	2
110	Ch	Pan	Scléro	*Oplismenus burmanni* (Retz.) P.Beauv.	0	0	0	0	0	0	0	0	0	0	0	0	0	0	0	0	0	0	0	0	2	2	2	2	2	0	2	2	2	0	2	2	0	0	0	0	0	0
111	Ch	Pan	Scléro	*Oplismenus hirtellus* (L.) P.Beauv.	3	3	3	3	3	3	3	3	3	3	3	3	3	2	2	0	0	0	0	0	0	0	0	0	0	0	0	0	0	0	0	0	0	0	0	0	0	0
112	Th	Cos	Ballo	*Oxalis corniculata* L.	0	0	0	0	2	2	0	0	2	2	0	0	0	0	2	0	2	0	2	2	2	0	0	2	0	0	2	0	2	2	2	0	2	2	0	0	0	2
113	Th	Cos	Ballo	*Oxalis latifolia* Kunth	2	5	3	3	3	4	4	3	3	2	2	0	0	0	0	0	0	0	0	0	0	0	0	0	0	0	0	0	0	0	0	0	0	0	0	0	0	0
114	Hc	Pan	Scléro	*Panicum humile* Nees ex Steudel	0	0	0	0	0	0	0	0	0	0	0	0	0	2	2	2	2	2	2	0	0	2	2	2	0	0	0	0	0	0	0	0	0	0	0	0	0	0
115	Hc	Plur-Afr	Scléro	*Panicum maximum* Jacq.	0	0	0	0	0	0	0	0	0	0	2	2	3	3	2	0	0	0	0	0	0	0	0	0	0	0	0	0	0	0	0	0	0	0	0	0	0	0
116	Ge	Pan	Scléro	*Paspalum notatum* Flügge	3	0	0	0	0	0	0	0	0	0	0	0	0	0	0	0	0	0	0	0	0	0	0	0	0	0	0	0	0	2	0	0	0	0	0	0	3	0
117	Hc	Intr	Sarco	*Passiflora edulis* Sims	0	0	0	0	0	0	0	0	0	0	0	0	1	0	0	0	1	0	0	0	0	0	0	1	0	0	0	0	0	1	0	0	0	0	0	0	0	0
118	P	Pan	Sarco	*Paullinia pinnata* L.	0	0	0	0	0	0	0	0	0	0	0	0	0	0	0	0	2	2	2	2	2	0	2	0	0	2	0	0	2	3	0	2	2	0	0	0	0	0
119	P	SZ(Z)	Sarco	*Pavetta virungensis* Bremek.	0	0	0	0	0	0	0	0	0	0	0	0	0	0	0	0	2	2	2	0	0	0	0	2	0	0	2	0	2	3	3	2	0	2	2	2	2	0
120	Th	Pan		*Pennisetum clandestinum* Hochst. ex Chiov.	4	0	0	0	0	2	3	0	0	0	0	0	0	2	2	2	2	0	0	2	0	2	0	0	0	2	0	2	2	0	2	0	3	3	0	2	2	2
121	Hc	Afr-Trop	Pogo	*Pennisetum purpureum* Schumach.	0	0	0	0	0	0	0	0	0	0	0	0	2	2	0	0	2	2	0	0	0	0	2	2	2	0	0	0	2	2	2	0	0	0	0	2	2	0
122	Hc	Pan	Scléro	*Pennisetum setaceum* (Forsk.) Chiov Grass	0	0	0	0	0	0	0	0	0	0	3	2	3	0	0	2	2	0	0	0	0	0	0	0	0	0	0	0	0	0	0	0	0	0	0	0	0	0
123	Hc	Mont	Scléro	*Pennisetum trachyphyllum* Pilg.	3	3	4	4	3	2	3	3	3	3	3	3	4	2	2	3	4	4	3	3	3	0	3	3	3	2	3	3	3	3	3	3	3	3	3	3	3	3
124	P	Afr-Trop		*Peponium vogelii* (Hook. f.) Engl.	1	1	1	0	0	1	1	0	1	0	0	0	0	0	0	0	0	0	0	0	0	0	0	0	0	0	0	0	0	0	0	0	0	0	0	0	0	0
125	P	Plur-Afr	Sarco	*Phoenix reclinata* Jacq.	0	0	0	0	0	0	0	0	0	0	0	0	2	0	0	0	0	0	0	0	0	0	0	0	0	0	0	0	0	1	0	0	0	0	0	0	0	0
126	Ge	Plur Afr	Scléro	*Phrahmites mauritianus* Kunth.	0	0	0	0	0	0	0	0	3	0	3	0	3	4	0	0	0	0	0	0	3	0	3	0	3	0	0	0	0	0	0	0	0	0	0	0	5	4
127	Th	Pan		*Phyllanthus amarus* Shumach. et Thonn.	2	2	0	2	0	2	2	0	2	0	0	0	2	0	0	0	0	2	0	0	2	0	0	0	0	2	0	2	0	2	2	0	0	0	0	2	2	2
128	Ch	SZ(OZ)	Scléro	*Plectranthus bojeri* (Benth.) Hedge	0	0	2	3	0	0	3	0	0	2	2	0	0	0	2	0	0	2	0	0	0	0	0	0	0	2	0	0	0	0	2	0	0	0	0	0	2	0
129				*Pneumatopteris blastophora* (Alston) Holtt.	0	0	0	0	2	0	0	0	0	0	0	0	0	0	0	0	3	0	2	0	2	0	0	2	0	2	2	0	0	0	0	0	0	0	0	0	0	0
130	Ch	Pal		*polygonum glabrum* Willd.	3	3	2	2	3	2	2	2	2	3	3	3	2	3	3	3	2	3	3	0	3	3	3	3	3	3	3	3	2	3	2	3	2	3	2	2	2	2
131	P	Afr-Am		*Psidium guajava* L.	0	0	0	0	0	0	0	0	0	0	0	0	0	0	0	0	1	0	0	0	0	0	0	0	0	0	0	0	0	2	0	0	0	0	0	0	1	0
132	Ch	Mont	Scléro	*Pycnostachys erici-rosenii* R.E.Fr.	0	0	0	0	0	0	0	0	0	0	0	2	2	1	2	1	1	0	0	0	0	0	0	0	0	0	0	0	0	0	0	0	0	0	0	0	0	0
133	P	SZ	Sarco	*Rhus vulgaris* Meikle	0	0	0	0	0	0	0	0	0	0	0	0	0	0	0	0	3	3	2	2	2	2	0	2	2	3	3	0	3	3	3	3	2	2	0	2	2	2
134	P	Cos		*Ricinus communis* L.	1	0	0	0	0	0	0	0	0	0	0	0	1	0	0	0	0	0	0	0	0	0	0	0	0	0	0	0	0	0	0	0	0	0	0	0	1	0
135				*Rubus pinnatus* Wild.	0	0	0	2	2	2	2	0	2	0	0	2	2	2	0	2	2	2	0	2	0	2	2	0	0	0	2	0	2	2	2	0	2	0	2	0	2	2
136	Ch	SZ(OZ)	Ptéro	*Rumex usambarensis* (Engl.) Dammer.	2	2	2	0	2	2	0	0	2	0	0	0	2	2	0	0	2	0	0	2	0	0	0	2	0	2	0	2	0	0	0	0	2	0	2	0	2	0
137				*Rumex beguaertii* De Wild.	2	2	0	0	0	0	0	0	0	0	0	2	2	0	0	2	2	0	0	0	0	0	0	0	0	0	0	0	0	0	0	0	0	0	0	0	2	0
138	Ch	SZ(OZ)	Pogo	*Sesamum angolense* Welw.	0	0	0	0	0	0	0	0	0	0	0	0	2	0	2	0	2	2	0	0	0	0	2	0	0	0	0	2	2	3	3	0	0	3	0	0	3	3
139	Th	Subcos	Scléro	*Setaria pumila* (Poir.) Roem. & Schult.	0	0	0	2	2	2	0	2	0	2	0	3	3	0	3	3	3	3	3	0	3	2	2	0	0	2	2	0	2	0	2	0	2	0	0	0	0	0
140	Hc	SZ(O)	Scléro	*Setaria kagerensis* Mez	0	0	0	0	0	0	0	0	0	0	0	0	0	3	0	2	2	2	2	2	0	0	0	0	0	0	0	0	0	0	0	0	0	0	0	0	0	0
141	Th	Pan	Desmo	*Sida acuta* Burm. f.	2	2	0	2	2	2	0	2	0	2	2	0	2	2	2	0	2	2	2	2	0	0	2	2	2	2	2	2	0	2	0	0	2	0	2	0	0	2
142	Th	Pan	Desmo	*Sida cordifolia* L.	0	0	2	0	0	2	0	0	2	0	0	0	0	0	1	0	0	0	1	0	0	0	2	2	0	0	0	0	2	2	2	0	0	0	2	2	2	0
143	Th	Pan	Desmo	*Sida rhombifolia* L.	0	0	0	0	0	0	0	0	0	0	2	0	0	0	0	2	0	0	0	0	2	0	0	0	0	2	0	0	0	2	0	2	0	2	0	0	0	0
144	Th	Pan	Desmo	*Sida veronicaefolia* Cav.	0	0	0	0	0	0	0	1	0	0	0	0	0	0	0	0	0	0	0	0	0	1	0	0	0	0	1	0	0	0	0	0	0	0	0	0	0	0
145	P	Plur-Afr	Sarco	*Smilax anceps* Willd.	0	0	0	0	0	0	0	0	0	0	0	0	0	0	0	0	3	2	3	2	3	0	0	3	0	2	2	0	3	3	2	2	2	2	0	2	0	2
146	Th	Pan	Sarco	*Solanum aculeastrum* Dunal.	0	0	0	0	0	0	0	0	2	0	0	0	0	0	0	0	2	0	0	0	2	0	0	0	0	0	0	0	0	2	0	0	0	0	0	0	0	2
147	Th	Cos	Sarco	*Solanum nigrum* L	0	0	0	0	0	0	0	0	0	0	0	0	0	0	0	0	0	0	1	1	0	0	0	0	0	0	0	1	1	0	0	0	0	0	1	1	1	1
148	Ch	Mont (EA)		*Sonchus luxurians* (R.E.Fr.) C.Jeffrey.	0	0	0	0	0	0	0	0	0	0	0	0	2	2	0	0	2	2	0	2	0	2	0	2	0	2	0	0	2	0	2	2	0	0	2	0	2	2
149				*Sphaeranthus suaveolens* (Forssk.) DC.	3	3	3	2	2	0	0	0	0	0	0	0	0	0	0	0	0	0	0	0	0	0	0	0	0	0	0	0	0	0	0	0	0	0	0	0	0	0

151	P	Pal	Sarco	*Stephania abyssinica* (Dill. & A. Rich.) Walp.	0	0	2	2	0	3	3	0	3	0	3	0	2	2	0	2	3	3	0	0	3	0	0	0	2	2	2	0	2	0	3	0	2	2	0	0	0	0	2	2	3
152	P	Afr-Trop	Ballo	*Sterculia tragacantha* Lindl.	0	0	0	0	0	0	0	0	0	0	0	0	0	0	0	3	3	3	2	0	3	2	0	2	0	2	2	3	0	3	0	3	3	0	2	2	2	2	2	2	2
153	Ch	Pal	Pogo	*Tagetes minuta* L.	3	3	0	2	2	0	2	0	4	4	4	2	2	0	4	2	2	0	4	2	0	4	2	4	2	4	0	0	4	4	0	2	0	0	4	0	4	3	0	2	3
154	Th	SZ	Ballo	*Tephrosia nana* Kotschy ex Schweinf.	0	0	0	2	0	0	0	2	0	2	2	0	0	2	0	0	0	0	0	0	0	0	0	0	0	0	0	0	0	0	0	0	0	0	0	0	0	0	0	0	0
155	Th	SZ(EOZ)	Ballo	*Trifolium repens* L.	3	3	0	3	3	0	3	0	3	0	0	0	0	0	0	0	0	0	0	0	0	0	0	0	0	0	0	0	0	0	0	0	0	0	0	0	0	0	0	0	0
156	He	Afr Am		*Tripsacum laxum* Nash.	2	0	0	0	0	0	0	0	0	0	0	0	3	4	2	2	0	3	0	3	3	0	3	0	3	0	0	3	0	3	0	0	2	0	3	0	2	3	0	0	0
157	Th	Pan	Desmo	*Triumfetta rhomboidea* Jacq.	2	2	1	2	1	2	2	2	2	2	2	2	2	2	2	2	2	2	2	2	2	2	2	2	2	2	2	2	2	2	2	2	2	2	2	2	2	2	2	2	2
158	P	Afr Trop	Sarco	*Uvaria angolensis* Welw. ex Oliv.	0	0	0	3	3	3	0	3	3	0	3	2	3	3	2	0	0	2	3	3	2	3	3	0	3	0	3	0	2	2	0	2	3	2	2	2	2	2	2	0	2
159	Ch	Mont	Ballo	*Virectaria major (K.Schum.) Verdc. subsp. major*	0	0	0	0	0	0	0	0	0	0	0	0	0	0	0	0	0	0	0	0	0	0	0	0	0	0	1	0	1	1	1	1	2	2	2	0	2	2	0	0	0
160	P	LSZ-Mo	Pogo	*Vernonia lasiopus* O. Hoffm.	2	0	1	0	1	2	2	1	0	1	0	0	0	0	0	0	0	0	0	0	0	0	0	0	0	0	0	0	0	0	0	0	0	0	0	0	0	0	0	0	0
161	P	Afro trop		*Wahlenbergia pulchella* Thulin	1	0	0	0	0	0	0	0	0	0	0	0	0	0	0	0	0	0	0	1	0	0	1	0	0	1	0	0	0	0	0	0	1	0	0	0	0	0	0	1	1
162	Th	Cos		*Xanthium strumarium* L.	3	3	3	0	0	0	0	0	0	0	0	0	0	0	0	0	0	0	0	0	0	0	0	0	0	2	0	0	2	2	2	0	0	3	3	3	3	2	2	2	4

Appendix 2. List of species collected

Families		Species		Common names
1	Acanthaceae	1	*Acanthus polystachyus* Del.	Itovu / igitovu
		2	*Hygrophila auriculata* (Schumach.) *Heine*	Bugannga
		3	*Hypoestes triflora* (Forssk.) Roem. & Schult.	Bukikiri
		4	*Justicia heterocarpa* T.Anderson.	umucaca
2	Alismataceae	5	*Burnatia enneandra* (Hochst.) Micheli	
3	Amaranthaceae	6	*Achyranthes aspera L*	Icaruza
		7	*Amaranthus viridis L*	Inyabutongo
4	Anacardiaceae	8	*Rhus vulgaris Meikle*	Umusagara
5	Annonaceae	9	*Uvaria angolensis* Welw. ex Oliv.	Umugimbu / Imihwi yo mw'ishamba
6	Apiaceae	10	*Centella asiatica* (L.) Urb.	Gutwikumwe
		11	*Steganotaenia araliacea* Hochst.	umuganasha
7	Apocynaceae	12	*Carissa spinarum L*	Umunyonza
		13	*Cynanchum schistoglossum* Schlt	
8	Arecaceae	14	*Phoenix reclinata* Jacq.	Igisandasanda
9	Aspleniaceae	15	*Asplenium onopteris* L.	Iraba
10	Asteraceae	16	*Acanthospermum austral* (Loefl.) Kuntze	Mwimbuyentaraza
			Ageratum conyzoides L	Akarura
		17	*Aspilia pluriseta* Schweinf.	Umwumira / icumwa
		18	*Bidens pilosa L*	Icanda
		19	*Blumea brevipens* (Oliv. & Hiern) *Wild*	Itabi ry'imbwa
		20	*Conyza aegyptiaca* (L.) Ait.	Mukobwandagowe
		21	*Crassocephalum crepidioides* (Benth.) S. Moore	Akaziraruguma
		22	*Crassocephalum montuosum* bumbense S. Moore	Igifurifuri
		23	*Crassocephalum MulticorymbosumS.*Moore	Igifurifuri
		24	*Crassocephalum vitellinum* (Benth.) S. Moore.	Umuyungubira
		25	*Dichrocephala bicolor* (Roth) Schltdl.	Umutambasha / umubuza
		26	*Emilia caespitosa Oliv.*	Akaryankwavu
		27	*Erlangea spissa* S. *Moore*	Umubebe
		28	*Galinsoga quadriradiata* Ruiz & Pav.	Akaryankwavu

		29	*Galinsonga parviflora* Cav.	Agakurasuka / kurisuka
		30	*Guizotia scabra* (Vis.) Chiov.	Ikizimyamuriro
		31	*Helichrysum keilii Moeser*	Manayeze / Akanyunga
		32	*Helichrysum odoratissimum* (L.) Sweet.	Isanganingoyi / Manayeze
		33	*Lactuca serriola* L.	
		34	*Microglossa pyrifolia* (Lam.) *Kuntze*	Umuhe
		35	*Sonchus luxurians* (R.E.Fr.) C.Jeffrey.	Akaziraruguma
		36	*Sphaeranthus suaveolens* (Forssk.) DC.	Akamazi / Ikinini
		37	*Tagetes minuta* L.	Ikimogimogi / sumurenga
		38	*Vernonia lasiopus* O. Hoffm.	Umukuraza
		39	*Xanthium strumarium* L.	
11	Boraginaceae	40	*Cordia africana* Lam.	Umuvugangoma
12	Brassicaceae	41	*Capsella bursa-pastoris* (L.) Medik.	Ubugurube
13	Campanulaceae	42	*Wahlenbergia pulchella Thulin*	Umurandaranda
14	Capparaceae	43	*Gynandropsis gynandra* (L.) Briq.	Isogi
15	Caryophyllaceae	44	*Drymaria cordata* (L.) Willd. ex Roem. &Schult.	Ikiracinzovu
16	Celastraceae	45	*Maytenus arbutifolia* (Hochst. ex A.Rich.) R.Wilczek	Umugunguma
17	Chenopodiaceae	46	*Chenopodium ugandae(Aellen) Aellen*	Umugombe
18	Combretaceae	47	*Combretum collinum* Fresen.	Umukoyoyo
19	Commelinaceae	48	*Commelina africana L*	Igiteza / Inteza
		49	*Commelina benghalensis* L	Igiteza / Inteza
20	Convolvulaceae	50	*Ipomea cairica* (L.) Sweet	Umumanuka
		51	*Ipomea involucrata*	Umuryanyoni
21	Crassulaceae	52	*Kalanchoe beniensis* De Wild.	Ikinetenete
22	Cucurbitaceae	53	*Momordica foetida* Schumach. & Thonn.	Umwishwa
		54	*Peponium vogellii* (Hook. f.) Engl.	Umutangatanga
23	Cyperaceae	55	*Bulbostylis densa* (Wall.) Hand.-Mazz.	Utunimbonimbo
		56	*Cyperus articulatus L*	Ubusa
		57	*Cyperus cyperoides* (L.) *Kuntze*	Inimbo
		58	*Cyperus distans* Linnaeus f.	Umurago / intaretare

		59	*Cyperus latifolius* Poir.	Igikangaga
		60	*Cyperus papyrus L*	Urufunzo / umuhotora
		61	*Kyllinga erecta* Schumach.	Ikija c'inimbo
		62	*Mariscus sumatrensis* (Retz.) J. Raynal.	Inimbo
24	Dioscoreaceae	63	*Dioscorea alata* L.	Amatuguy'imfyisi
25	Dracaenaceae	64	*Dracaena steudneri* Engl.	Igitongati
26	Euphorbiaceae	65	*Euphorbia heterophylla* L.	Igicamate
		66	*Euphorbia hirta* L.	Akanyaruguma
		67	*Euphorbia tirucalli* L.	Umunyari
		68	*Phyllanthus amarus* Shumach. et Thonn.	
		69	*Ricinus communis* L.	Ikinobonobo
27	Fabaceae	70	*Acacia polyacantha* Willd.	Umugunga /Umusange
		71	*Acacia sieberana* DC.	Umunyinya
		72	*Albizia gummifera* (J.F.Gmel.) C.A.Sm.	Umusebeyi/umusaravyondo
		73	*Cassia corymbosa* Lam.	
		74	*Cassia didymobotrya* Fres.	Umubagabaga
		75	*Cassia occidentalis* L.	Umuyokayoka
		76	*Crotalaria pallida Aiton*	Ikinyenzogera
		77	*Crotalaria pumila* Ort. Chipil.	Akanyenzogera
		78	*Crotalaria retusa* L.	Akanyenzogera
		79	*Eriosema montanum* Baker f.	Umupfunyantoke
		80	*Indigofera spicata* Forssk.	
		81	*Kotschya strigosa* (Benth.) Dewit & P.A.Duvign.	Umushiha / umucutsa
		82	*Mimosa diplotricha* C.Wright.	Ubuyabu / Imigeyogeyo
		83	*Mimosa pigra L*	Ubuyabu / Imigeyogeyo
		84	*Neorautanenia mitis* (A.Rich.) Verdc.	Igitembetembe/umuguhaguha
		85	*Tephrosia nana* Kotschy ex Schweinf.	Ntibuhunwa
		86	*Trifolium repens* L.	
28	Hypericaceae	87	*Harungana madagascariensis* Lam. & *Poir*	Umushayishayi
29	Lamiaceae	88	*Leonotis nepetifolia* (L.) R. Br.	Umutongotongo
		89	*Leucas martinicencis* (Jacq.) R. Br.	Akanyamapfundo
		90	*Plectranthus bojeri* (Benth.) Hedge	Manyama
		91	*Pycnostachys erici-rosenii* R.E.Fr.	Umutsinduka
	Malvaceae	92	*Hibiscus fuscus Garcke*	Umutete

30		93	*Hibiscus noldeae* Baker f.	Umuvumvu
		94	*Sida acuta* Burm. f.	Umuvumvu
		95	*Sida cordifolia* L.	Umuvumvurweru
		96	*Sida rhombifolia* L.	Akavumvu
		97	*Sida veronicaefolia* Cav.	
31	Melastomataceae	98	*Dissotis trothae* Gilg.	Umushonge(sha)
32	Menispermaceae	99	*Cissampelos mucronataA*.Rich.	Umuhanda
		100	*Stephania abyssinica* (Dill. & A. Rich.) Walp.	Umuhanda
33	Moraceae	101	*Ficus vallis-choudae* L.	Igikuyu
		102	*Morus indica* L.	
34	Myrsinaceae	103	*Maesa lanceolata Forsskal*	Umuhangahanga
35	Myrtaceae	104	*Psidium guajava* L.	Ipera
36	Nephrolepidaceae	105	*Nephrolepis undulata* (Afzel. ex Sw.) J. Sm.	Ubuhumbirajana
37	Oleandraceae	106	*Arthropteris anniana* Lawalrée	
		107	*Arthropteris orientalis* (J.F. Gmel.)	Udushusrushuru
38	Onagraceae	108	*Ludwigia abyssinica* A. Rich.	
39	Oxalidaceae	109	*Biophytum helenae* Buscal. & Muschl.	Tinyabakwe
		110	*Oxalis corniculata* L.	Umunyuwanyamanza
		111	*Oxalis latifolia Kunth*	Ingonga
40	Passifloraceae	112	*Passiflora edulis Sims*	Amabungo
41	Pedaliaceae	113	*Sesamum angolense* Welw.	Umurendarenda/umukakambari
42	Poaceae	114	*Aristida adoensis* Hochst. ex A.Rich.	Agatsindampfizi
		115	*Arundinalia alpina* K.Schum.	Umugano
		116	*Cenchrus unisetus (Nees)* (Nees) *Morrone*	Umukenke
		117	*Cymbopogon giganteus* Chiov.	igikenkekenke
		118	*Cynodon dactylon* (L.) Pers.	Urucaca
		119	*Cynodon nemfluensis* Vanderyst .	Urucaca

		120	*Digitaria abyssinica* (Hochst. ex A.Rich.) *Stapf*	Urwiri
		121	*Digitaria horizontalis* Willd.	Urwiri
		122	*Eleusine indica* (L.) Gaertn.	Urwamfu
		123	*Eragrostis olivacea* K.Schum.	Ishinge
		124	*Eragrostis tenuifolia*(A.Rich.) Hochst. ex Steud.	Ubusuga
		125	*Hyparrhenia dichroa* (Steud.) Stapf.	igisekenkanya
		126	*Imperata cylindrica* (L.) Beauv.	Isovu
		127	*Melinis minutiflora* P.Beauv.	Ikinyamavuta
		128	*Melinis repens* (Willd.) *Zizka*	Urwarikafundi
		129	*Oplismenus burmanni* (Retz.) P.Beauv.	Igona / igono
		130	*Oplismenus hirtellus* (L.) P.Beauv.	Igona / igono
		131	*Panicum humile* Nees ex Steudel	
		132	*Panicum maximum* Jacq.	Ikinywabuki / integarubingo
		133	*Paspalum notatum Flügge*	Akanyatsi
		134	*Pennisetum clandestinum* Hochst. ex Chiov.	
		135	*Pennisetum purpureum* Schumach.	Ikibingo
		136	*Pennisetum setaceum* (Forsk.) Chiov Grass	
		137	*Pennisetum trachyphyllum* Pilg.	Igikaranka
		138	*Phrahmites mauritianus* Kunth.	Amatete / amarenga
		139	*Setaria pumila* (Poir.) Roem. & Schult.	Isheshe
		140	*Setaria kagerensis Mez*	Igikaranka
		141	*Tripsacum laxum* Nash.	
43	Polygonaceae	142	*Polygonum glabrum* Willd.	
		143	*Rumex usambarensis* (Engl.) Dammer.	Umufumbegeti

		144	*Rumex beguaertii* De Wild.	Isesabirego
44	Rhamnaceae	145	*Maesopsis eminii* Engl.	Umuhumuza / indunga
45	Rosaceae	146	*Rubus pinnatus* Wild.	Inkerere
46	Rubiaceae	147	*Hymenodictyon floribundum* (Hochst. & Steud.) B.L.Rob	
		148	*Pavetta virungensis* Bremek.	Umunyamabuye
		149	*Virectaria major* (K.Schum.) Verdc. subsp. major	Umukizikizi
47	Sapindaceae	150	*Paullinia pinnata* L.	Umusarara/ umusarasara
48	Smilacaceae	151	*Smilax anceps* Willd.	Umusuri, Umurerajuru
49	Solanaceae	152	*Datura stramonium* L.	Intibwa
		153	*Solanum aculeastrum* Dunal.	Umutobotobo
		154	*Solanum nigrum L*	Isogo
50	Sterculiaceae	155	*Dombeya buettneri K.*Schum.	Umukongwa
		156	*Sterculia tragacantha* Lindl.	Igitakataka / umukungwe
51	Thelypteridaceae	157	*Pneumatopteris blastophora* (Alston) Holtt.	
52	Tiliaceae	158	*Triumfetta rhomboidea* Jacq. / Urena lobata L.	Umukururantama
53	Verbenaceae	159	*Clerodendrum johnstonii* Oliv.	Ikiziranyenzi
		160	*Lantana camara* L.	Umuhengerihengeri
54	Vitaceae	161	*Cyphostemma adenocaule* (Steud. ex A. Rich.)	Akaboza / Umubombobombo

Printed by Books on Demand GmbH, Norderstedt / Germany